AF685176

Practical Interventional Radiology of the Musculoskeletal System

LANDERS

PRACTICAL

Interventional Radiology of the Musculoskeletal System

edited by

DAVID J. WILSON, MBBS, BSc, MRCP, FRCR

Consultant Radiologist, Nuffield Orthopaedic Centre, and Clinical Lecturer, University of Oxford, Oxford, UK

Edward Arnold

A member of the Hodder Headline Group

LONDON BOSTON SYDNEY AUCKLAND

First published in Great Britain 1995 by
Edward Arnold, a division of Hodder Headline PLC,
338 Euston Road, London NW1 3BH

Distributed in the Americas by
Little Brown and Company,
34 Beacon Street, Boston MA 02108

© 1995 Edward Arnold

All rights reserved. No part of this publication may be reproduced or transmitted in any form or by any means, electronically or mechanically, including photocopying, recording or any information storage or retrieval system, without either prior permission in writing from the publisher or a license permitting restricted copying. In the United Kingdom such licences are issued by the Copyright Licensing Agency: 90 Tottenham Court Road, London W1P 9HE.

Whilst the advice and information in this book is believed to be true and accurate at the date of going to press, neither the authors nor the publisher can accept any legal responsibility or liability for any errors or omissions that may be made. In particular (but without limiting the generality of the preceding disclaimer) every effort has been made to check drug dosages; however, it is still possible that errors have been missed. Furthermore, dosage schedules are constantly being revised and new side effects recognized. For these reasons the reader is strongly urged to consult the drug companies' printed instructions before administering any of the drugs recommended in this book.

British Library Cataloguing in Publication Data
A catalogue record for this book is available from the British Library

Library of Congress Cataloging-in-Publication Data
A catalog record for this book is available from the Library of Congress

ISBN 0 340 57645 6

1 2 3 4 5 95 96 97 98 99

Typeset in 11/12 Bembo by
Textype Typesetters
Printed and bound in Great Britain by
The Bath Press, Avon

Contents

Preface

What is interventional radiology? It certainly should include procedures that employ imaging whilst treating disease. But that definition is too simple: most of us would include percutaneous biopsy as an interventional technique and, although often an essential prerequisite to treatment, it certainly never makes a patient feel better. Similarly the surgeon may use fluoroscopy to guide placement of fixation devices and we do not regard this as a radiological technique. Perhaps it could be regarded as a diagnostic or therapeutic technique that requires informed consent and is used by radiologists to treat or diagnose disease. However, this definition would include arthrography under sedation and I felt that this would be too wide a remit for this book.

Puzzled by all of this, I asked a senior academic radiologist how he would define interventional radiology. 'Oh that's easy, it's any imaging technique that enables the radiologist to intervene between the surgeon and his wallet.'

With this pleasant thought in mind I have been fortunate in persuading a group of well known authors to contribute to this collection of essays on the techniques that skeletal radiologists employ. All radiologists may be regarded as interventionalists; those of us who take an interest in musculoskeletal disease would consider many of the following topics as routine work whilst others are on the frontiers of development. I hope that you find some of the new concepts useful in your interventional endeavours.

David J. Wilson
1995

Contributors

Charles Aprill MD, Magnolia Diagnostics Inc., New Orleans, Lousiana, USA

Nikolai Bogduk MB BS, PhD, Hon FACRM, Professor of Anatomy, Faculty of Medicine and Health Sciences, University of Newcastle, Newcastle, NSW, Australia

Claude Depriester MD, Unit of Neuroradiology, University Hospital, Amiens, France

Hervé Deramond MD, Unit of Neuroradiology, University Hospital, Amiens, France

Richard Derby MD, Spinecare, Daly City, California, USA

Pierre Galibert MD, Department of Neurosurgery, University Hospital, Amiens, France

Nagui S. T. Gendi MRCP, Senior Registrar in Rheumatology and Rehabiliation, The Nuffield Orthopaedic Centre, Oxford, UK

Philip Gishen MB, BCh, DMRD, FRCR, Consultant Radiologist, King's College Hospital, London, UK

James E. Jackson MRCP, FRCR, Senior Lecturer and Consultant Radiologist, Royal Postgraduate Medical School, Hammersmith Hospital, London, UK

Iain McCall MB, ChB, DMRD, FRCR, Consultant Radiologist, The Robert Jones and Agnes Hunt Orthopaedic Hospital, Oswestry, UK

John A. McCulloch MD, FRCS (C), Professor of Orthopaedics, Northwestern Ohio Universities College of Medicine, Ohio, USA

Gary Onik MD, Princetown Hospital, Orlando, USA

John Outhwaite MCRP, Consultant Physician in Rehabilitation Medicine, Nuffield Orthopaedic Centre, Oxford, UK

David J. Wilson MBBS, BSc, MRCP, FRCR, Consultant Radiologist, Nuffield Orthopaedic Centre, and Senior Clinical Lecturer, University of Oxford, Oxford, UK

Fiona M. Witham BSc AKC, MBBS, MRCP, FRCR, Senior Registrar, Department of Radiology, Southampton University Hospitals, Southampton, UK

1
JOINTS

CHAPTER 1

Joint aspiration

David J. Wilson

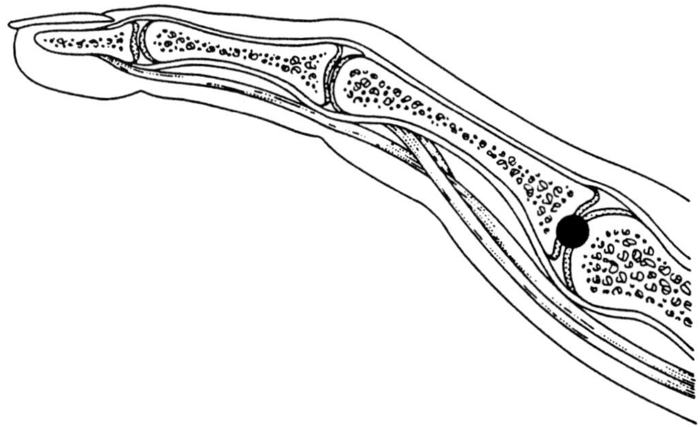

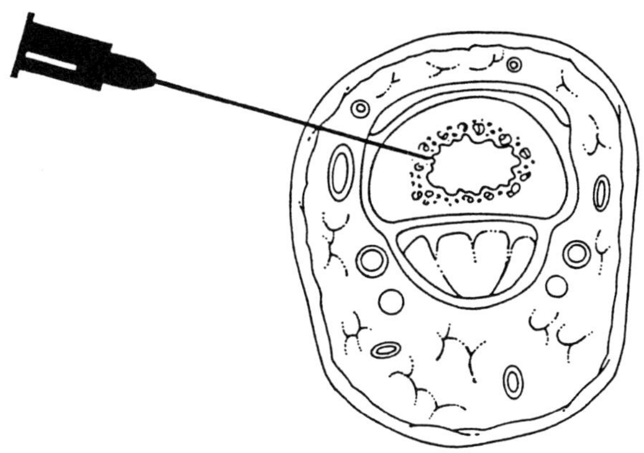

INTRODUCTION

The direction of needles into joints is an important clinical skill required by clinicians and radiologists who investigate and treat arthropathies.[1] A number of relatively superficial joints are amenable to puncture by palpation and visual guidance. If aspiration fails then imaging should be employed for a second attempt. Deeper joints will always require imaging to direct placement and confirm the needle's location.

Joints may be injected to introduce contrast agents for arthrography, to remove fluid to relieve symptoms and diagnose sepsis or crystal arthropathy. They may also be injected to introduce therapeutic agents including anaesthetics, steroids and radioactive substances.

GENERAL

Consent

It is not customary to require written consent for joint aspiration as the majority of procedures are akin to venepuncture and are performed without sedation. Clearly the radiologist will be required to obtain verbal consent in all cases and should be required to seek written consent when sedation is planned. There is some value in a brief written explanation given in advance, possibly at the time of appointment, as this seems to alleviate some of the anxiety generated by the unknown. The patient should be asked about allergies, current medication and a history of cardiac or respiratory disorders. The likely need for sedation may be judged during this consultation.

Indications for joint puncture

ARTHROGRAPHY

Although the techniques of needle puncture are the same, full discussion of this topic is beyond the scope of this chapter.

INFECTION

Aspiration to exclude infection is a common indication and should be undertaken with little hesitation when this diagnosis is considered. In older or immunosuppressed patients the symptoms and signs may be subtle or non-existent. It is important to look for crystals in purulent effusions particularly when the Gram stain and cultures are negative.[2] If the adjacent bone is affected then the histological findings from a core taken through the joint and into the bone are more likely to yield positive cultures and should give diagnostic histology even when the organism cannot be isolated. Proven infection should be treated by surgical drainage and intravenous antibiotics. There is no place for intra-articular antibiotics or repeated aspiration. Joint infections are a serious disorder and should be treated early and aggressively to avoid destruction and premature osteoarthritis.

RELIEF OF PRESSURE

Some authorities consider that the high intra-articular pressures that may occur in transient synovitis of the hip not only are the cause of pain but may lead to intra-osseous venous tamponade and secondary osteonecrosis. There is debate on the relationship between Legg–Calvé–Perthes disease and prior transient synovitis. It is reasonable to aspirate large and painful hip effusions to relieve pain and possibly to alleviate pressure.[3,4] The same considerations apply to large haemarthrosis in any joint and in addition there is value in reducing the quantity of blood, since blood may excite a reactive synovitis. It is probable that the aggressive synovitis seen in haemophilia and in the coagulopathies is the effect of blood in prolonged contact with the joint's surface. This hypothesis leads to the conclusion that blood should be removed from the joint space as a therapeutic measure. Certainly the relief of pressure and pain will allow early mobilization and less risk of stiffness and contractures.

DIAGNOSTIC BLOCK

In cases where there is doubt that the pain arises from the joint, the introduction of local anaesthetic may clinch a diagnosis.

THERAPEUTIC INJECTION

Injection of anaesthetic and steroid into a joint is popular as a therapeutic measure. Despite the risk of introducing infection and the potential for neuropathic type of damage in the period following injection, this technique has proved to be a reasonably harmless procedure. Where problems have ensued they are usually the result of multiple injections. It is wise to use only those agents recommended for intra-articular injection (e.g. triamcinolone acetonide, Kenalog–Squibb) and to take great care in excluding infection prior to treatment.

Synovitis may be treated by injection of radiopharmaceuticals (Chapter 2).

Guidance

Sadly, many so-called joint punctures achieved without guidance are likely to end in soft tissue adjacent to the joint space. A 'dry tap' cannot be confidently assumed unless there is some evidence that the needle was truly intra-articular. In practice this is best achieved by introducing a non-ionic water-soluble contrast agent and taking a plain radiograph or fluoroscopic photograph. Iodine concentrations of 180–300 mg/ml are suitable, using the higher concentrations for the smaller joints or where an effusion may dilute the injected material. Contrast agents have an antiseptic property and any specimen for culture should be obtained before injection. An alternative is to inject saline or air, which may then be seen by ultrasound, CT or MRI. This is of particular use when there is a history of allergy to iodine-related products.

Needles

Normal venepuncture needles are usually all that is required. The length will depend on the size of the patient and the joint involved. The hip and the shoulder are usually fairly deep and a lumbar puncture needle will be most suitable. Ideally the needle should have a trocar to prevent plugging with soft tissues, although this is not essential. A 21 G needle is appropriate except for the most glutinous of effusions, when a 19 G needle may be required. Always use sterile disposable needles; they are sharper and less likely to break.

Local anaesthetic

The experienced operator may be able to puncture a joint reliably with a single pass of the needle. In this case an anaesthetic is more likely to increase the discomfort during infiltration and may be omitted. Only a few have such skills and, if in doubt, infiltration of 1% lignocaine without adrenaline is a sensible preliminary procedure. Introducing some anaesthetic into the joint itself may alleviate some of the mild discomfort that follows the procedure. In children a local anaesthetic jelly applied to the area at least one and a half hours in advance is strongly recommended. It is best carefully to instruct nursing staff where to place the jelly as their concept of surface marking of the joint may be well adrift from the truth. A few patients exhibit a cutaneous allergic reaction to the jelly but the advantages far outweigh this small risk.

Sedation

Sedation is indicated only in particularly anxious patients or children, except in injection of the temporomandibular joint (see below). Intravenous midazolam (Hypnovel Versed Roche) or other benzodiazepines are suitable. Reversant drugs (flumazenil, Annexate Roche) and resuscitation facilities should always be available and a trained member of staff who is not performing the puncture should be to hand. If sedation is used, emergency anaesthetic assistance should always be available, normally via a crash bleep system (Royal College of Radiologists, Faculty of Anaesthetists working party in Radiology Department Sedation). It is desirable to monitor the sedated patient with a pulse oximeter. Fortunately, it is unlikely that joint puncture will require intravenous analgesia in adults. In infants and children anaesthetic advice should be sought prior to sedation as some will require general anaesthesia and sedation should not be undertaken lightly or by the inexperienced.

Complications

MECHANICAL DAMAGE

Vessels and nerves may be damaged leading to haemorrhage, aneurysm, paralysis, anaesthesia or neuroma. Fortunately these are very rare complications. The risk may be minimized by careful planning of injection routes, immediate withdrawal when there is acute pain and the use of thin guide needles prior to the introduction of larger biopsy devices. Damage to the articular surface is rare and not a serious risk.

INFECTION

Even with scrupulous aseptic technique infection remains a risk although it is much less common than aseptic or 'chemical' synovitis following injection of contrast agents or steroids. Any patient who suffers persisting discomfort, increased effusion or toxicity following joint aspiration should be urgently re-examined and a repeat aspiration undertaken if there is any doubt.

ALLERGY

Systemic reaction to contrast agents is much less likely following injection into joints than following intravenous administration. Despite this, it is mandatory to take a history of allergy to contrast agents and to have suitable resuscitation facilities to hand. Chemical synovitis and post-procedural pain are less frequent when non-ionic contrast agents are used.

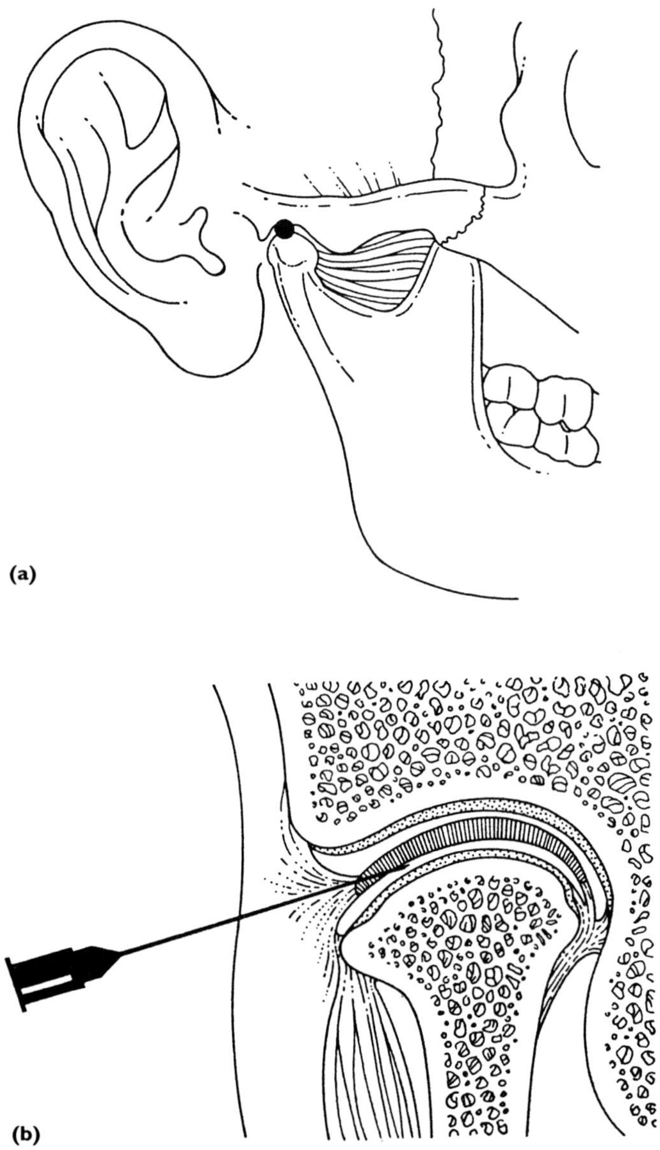

(a)

(b)

Fig. 1.1
Coronal and lateral views of a puncture of the temporomandibular joint. Try to avoid penetrating the upper joint space. The area is very vascular.

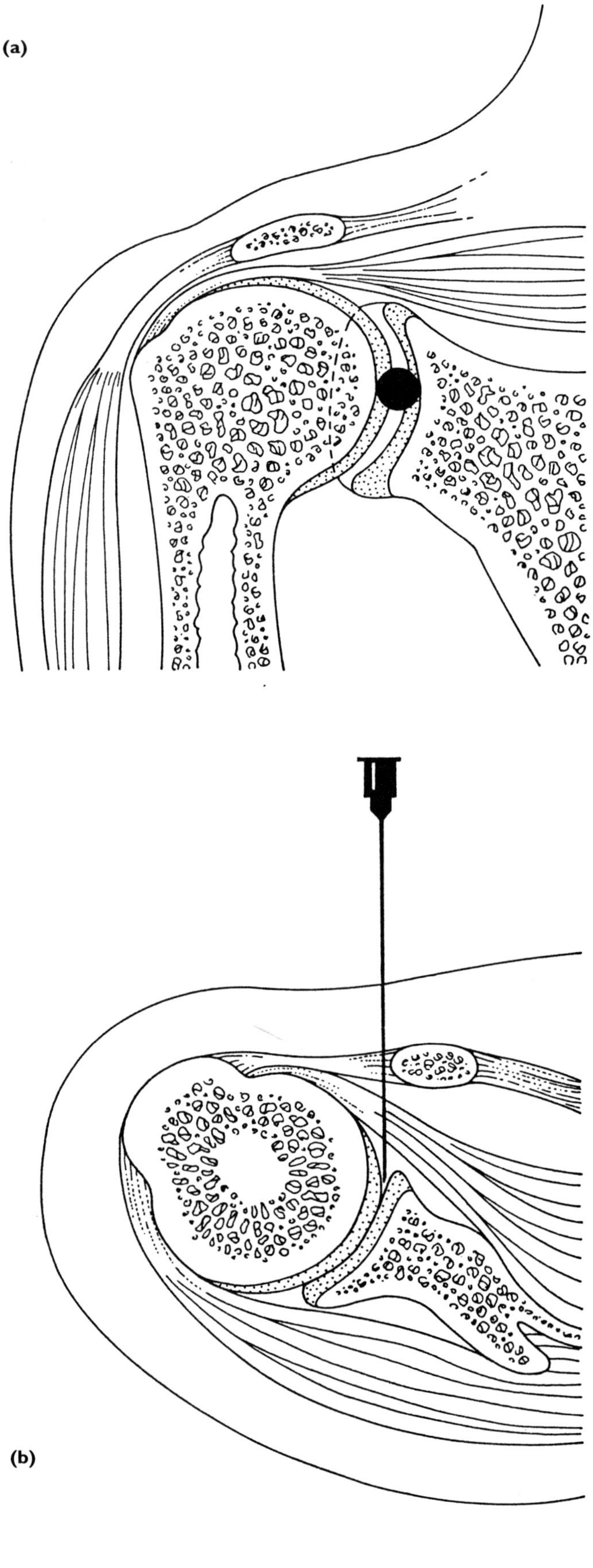

(a)

(b)

Fig. 1.2
Anterior approach to the glenohumeral joint. Note the obliquity of the glenoid.

SPECIFIC JOINTS

Temporomandibular joint

Puncture of the temporomandibular joint[5] (Fig. 1.1) is one of the more difficult procedures, requiring considerable experience and skill. The area is very sensitive and many prefer to sedate the patient as well as use local anaesthesia. Two to five milligrams of midazolam titrated against slurring of the patient's speech allows a 5- to10-minute period of co-operative but amnesic sedation. The joint normally contains two watertight spaces divided by a horizontal meniscus. Fluoroscopic guidance is highly desirable; puncture without guidance is possible but very unreliable. The apex of the mandibular condyle should be marked on the skin using fluoroscopy. It will be necessary to angle the beam cephalad or obliquely to project the joint away from the otherwise confusing images created by the overlying opposite joint. The direction of the needle will be along the resultant beam. This may be achieved by use of a 'C arm' fluoroscopy system or by inclining the patient's head over a pillow with the beam fixed in the vertical plane. Local anaesthetic is infiltrated and the needle directed to hit the apex of the condyle whilst the mouth is closed. The patient should be asked to open the mouth slowly. The needle tip is then 'walked' over the condyle to slide 1 or 2 mm past into the inferior joint space. Passing a few millimetres further, impacting the temporal fossa, would allow puncture of the superior joint space. The area is very vascular and bruising is common. Excessive use of anaesthetic may lead to temporary facial nerve paresis, requiring the eye to be protected by a patch until the anaesthesia has worn off.

Glenohumeral joint

The cup and ball shape of the glenohumeral joint should be considered when marking the skin (Fig. 1.2). A true radiological AP projection shows the apparent gap between glenoid and humerus but if the needle is introduced vertically it will hit the upturned glenoid brim and fail to enter the joint. The easiest approach is to lie the patient supine without rotation and mark the mid-point of the joint; Fig. 1.1 shows how the needle will then enter the joint if introduced vertical to the patient rather than the joint. Any deflection medially suggests that it has met the inner side of the glenoid brim and will need to be redirected. A slight give and a curve of the needle tip towards the humeral head are often observed in the last few millimetres. The tip of the needle should not be impacted into the glenoid cartilage as this will damage its surface and block flow.

Acromioclavicular joint and bursae

Injection into the acromioclavicular joints is rarely indicated. A combination of fluoroscopy and palpation facilitates puncture, which is best achieved in a vertical direction from above. An anterior approach is a useful alternative. The joint is very superficial and there are no special precautions.

The sub-acromial bursa is separate from the glenohumeral joint in most patients. Puncture is achieved by directing the needle anteriorly under the brim of the acromion. The most common indication is to perform arthrography to show partial thickness tears of the superior surface of the rotator cuff.

Elbow joint

In the case of the elbow joint (Fig. 1.3), a lateral approach is most comfortable for the patient and avoids the neurovascular bundle. The elbow is more comfortable when flexed. The co-operative patient may sit beside the fluoroscopy table although an assistant should ensure that the patient is supported should the injection lead to fainting. Lateral fluoroscopy should identify the radio-humeral joint and then a direct puncture is made into the space, which is only half a centimetre or so below the skin. This will avoid the radial nerve as it runs around the neck of the proximal radius. Because of the ball and socket nature of this part of the elbow the needle should be directed a little closer to the humeral side of the joint, then introduced with slight distal direction to avoid impact with the brim of the radius.

Wrist joints

The wrist (Fig. 1.4) consists of three major compartments. In young patients they are normally separate. Rupture between spaces is common in older individuals.

RADIO-ULNAR

The smallest space in the wrist is easily entered from a dorsal puncture with AP fluoroscopy. The needle will probably touch either the radius or the ulna. The joint space is around 1 cm in length and the needle should not be directed too proximal.

PROXIMAL CARPAL

The most comfortable approach is dorsal at the mid-point of the radio-scaphoid joint space with the wrist slightly flexed. This is a ball and cup joint and the needle should be directed a little proximal, towards the elbow. To achieve this it is best to enter the skin a little more

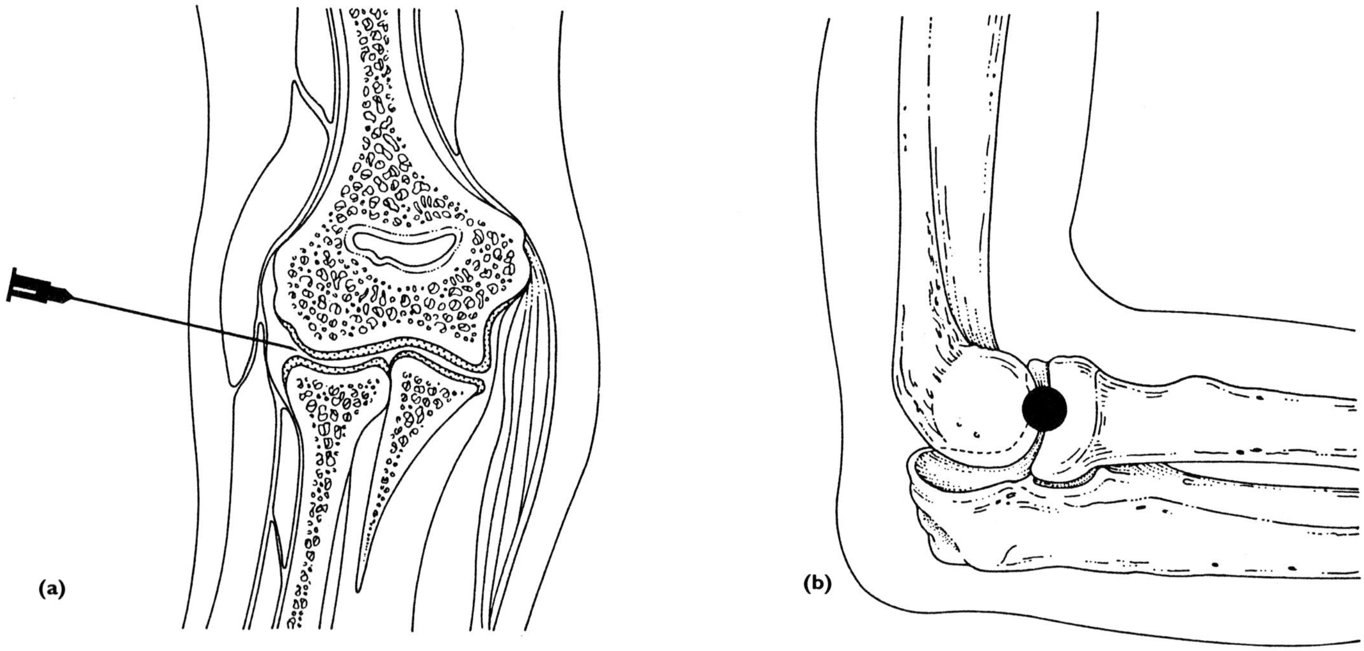

Fig. 1.3
The radio-humeral joint is very close to the skin. Fluoroscopic control helps to avoid the radial nerve.

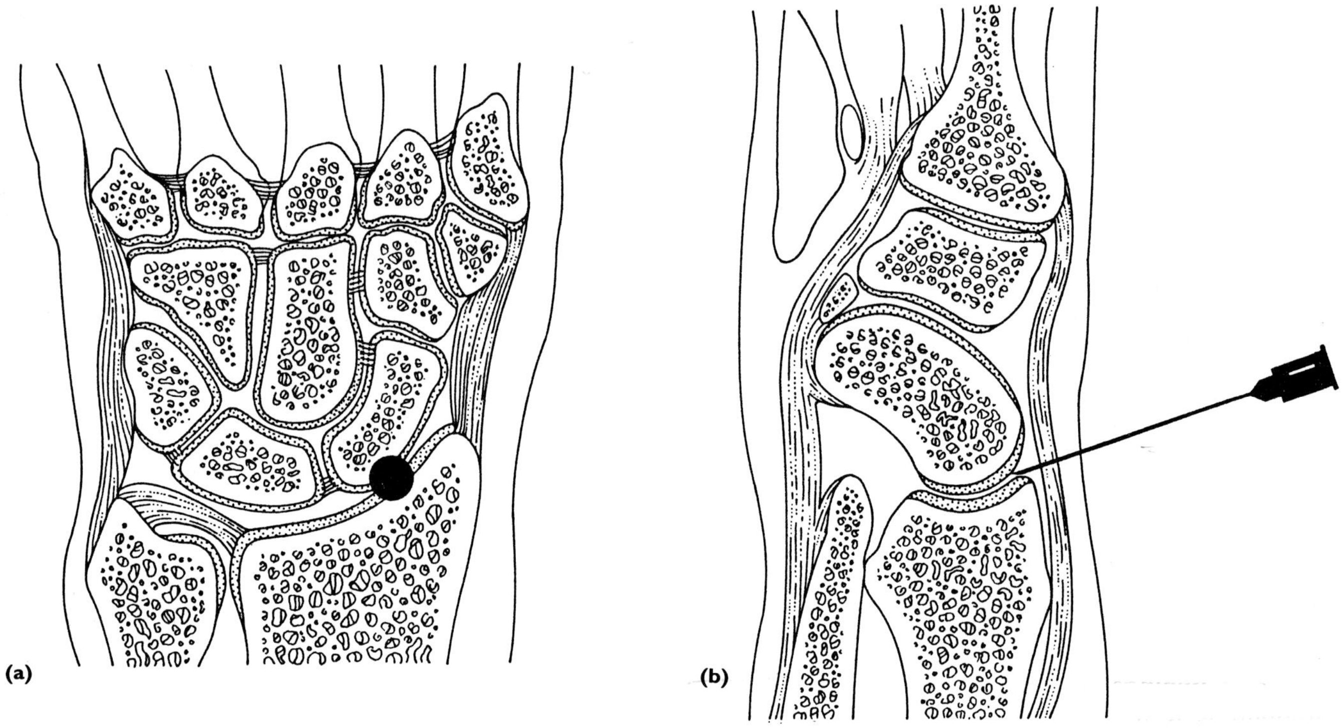

Fig. 1.4
The wrist is easily entered via a dorsal approach.

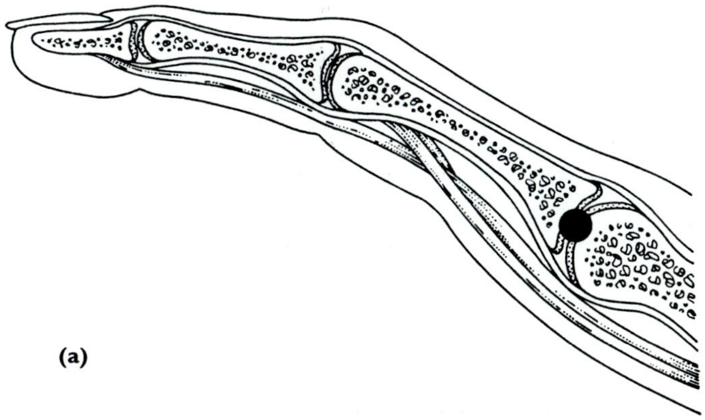
(a)

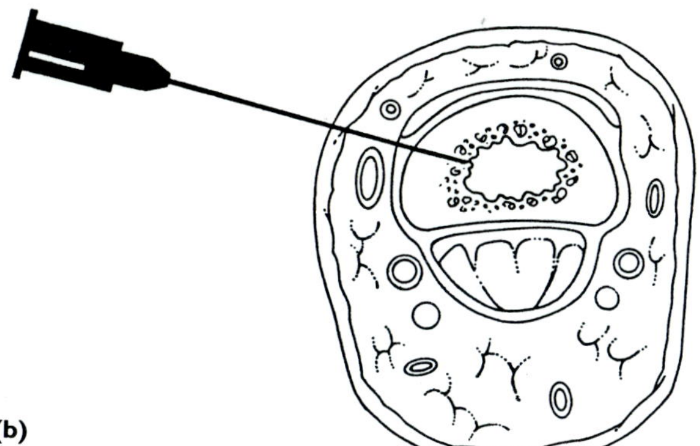
(b)

Fig. 1.5
Small but painful to puncture, the interphalangeal joints are bounded by vessels, nerves and tendons.

distal than the projection of the joint observed on fluoroscopy.

CARPAL

A direct dorsal puncture should be straightforward, selecting any of the intercarpal spaces.

Interphalangeal joints

A lateral injection into the interphalangeal joints (Fig. 1.5) may impinge on digital neurovascular structure although it is reasonably reliable. A dorsal approach penetrates tendons and would be painful. A dorsal oblique route is preferred.

Hip joint

Imaging guidance is mandatory in the hip joint (Fig 1.6), because of its depth. Several approaches may be advocated:

1 vertical anterior
2 direct lateral
3 abducted leg, medial
4 oblique anterior.

It is useful to be proficient in all these techniques as a difficult puncture may require an alternative approach. This is particularly true when there are osteophytes, previous surgery to the capsule or a dysplastic joint. The method that seems to give the most reliable result with the least discomfort is the anterior oblique method. It is a little more complex than the direct anterior approach but avoids the vascular bundle and minimizes the chance of hitting the acetabular brim before penetrating the joint. The skin entry is directly above the mid-trochanteric point. The needle is aimed obliquely towards the centre of the femoral head. As soon as bone is touched, the fluoroscopic image will show whether the needle is far enough along the femoral neck. If it is too proximal it should be withdrawn partially and redirected. It is rarely necessary to image in any other plane than anteriorly. If there is difficulty in penetrating the joint despite apparently good position then the leg should be voluntarily flexed at the hip whilst the knee is held down. This tenses the capsule and allows the needle to enter the joint.

The direct anterior and lateral approaches also require fluoroscopic guidance, preferably in more than one plane. The vascular bundle should be identified by palpation then marked on the skin. Both routes may impact the acetabular brim, and the relatively direct approach to the thin joint space may result in a needle half in and half out of the capsule.

Knee joint

Imaging is rarely required in order to enter the knee joint (Fig. 1.7). The easiest approach is laterally into the retropatellar space, although some argue that the medial retropatellar space is more open and less likely to be affected by osteophytes. The patella should be relaxed and mobile and then pulled towards the examiner. The needle is introduced behind the patella and allowed to move back with the other structures as pressure on the patella is released. It then seems to slide into the joint without probing, minimizing impact on the articular surface and reducing pain. An injection of fluid or air will cause the medial side of the joint to bulge, confirming correct placement. Flow should be free and the plunger of the syringe should not recoil.

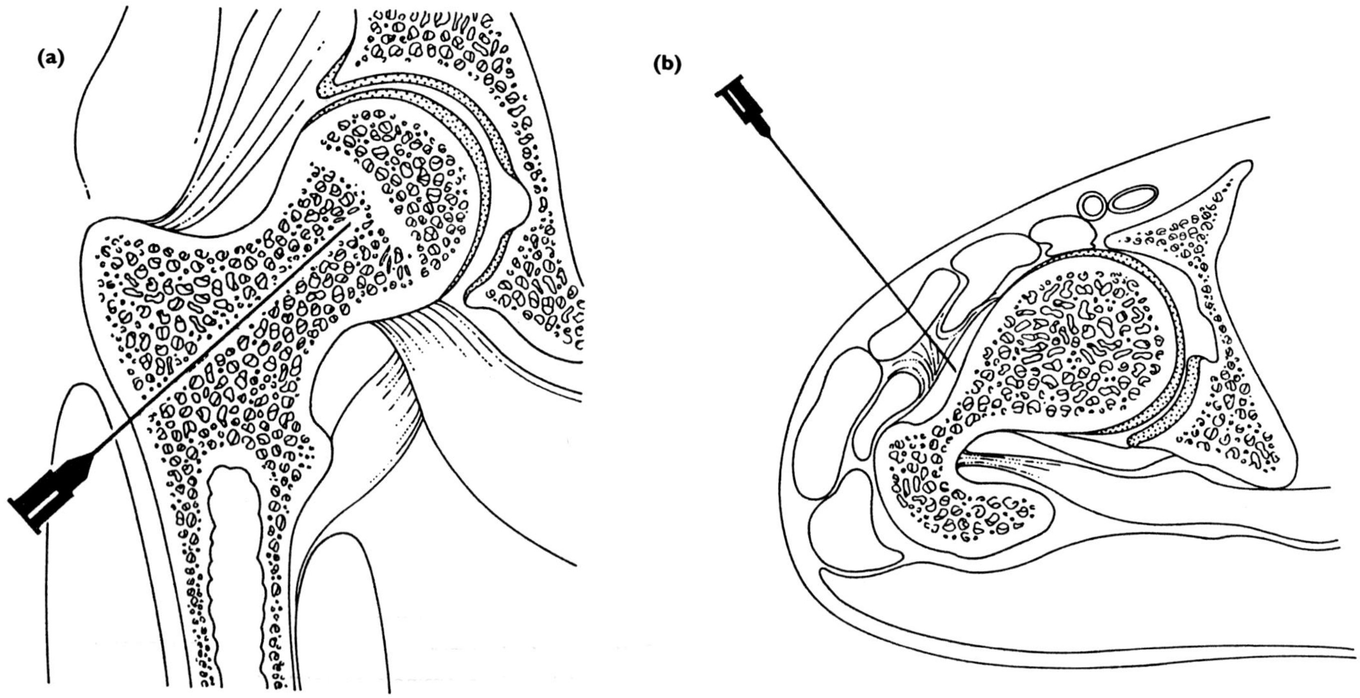

Fig. 1.6
For the hip joint, the antero-oblique approach is the route most likely to succeed.

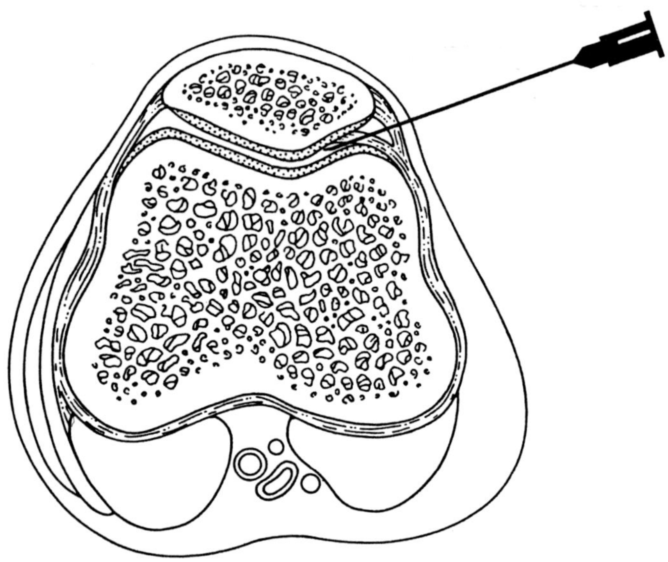

Fig. 1.7
Knee joints punctures rarely need imaging guidance.

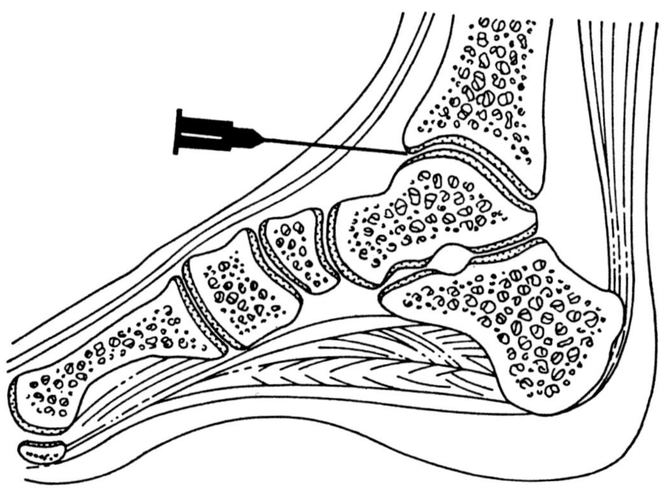

Fig. 1.8
The ankle joint should be punctured avoiding the dorsalis pedis artery and the extensor tendons. Palpate these structures beforehand.

Ankle joint

In the ankle, the mortice joint (Fig. 1.8) is best approached in the mid-line anteriorly with the needle directed a little cephalad. This avoids the brim of the anterior tibia. Extending the foot allows better access. Lateral fluoroscopy is ideal. Care should be taken to palpate and mark the dorsalis pedis artery, which is close to the skin entry point.

Subtalar joints

The subtalar joints would be difficult to enter were there not a normal connection with the talo-navicular joint. Puncture is therefore similar to the ankle joint in the mid-line anteriorly but this time directing the needle between talus and navicular more towards the toes.

Tarsal joints

Direct dorsal approach using AP fluoroscopy is best for the tarsal joints, which are rarely punctured.

Interphalangeal joints

The same considerations apply as for the fingers.

Sacroiliac joints

It is difficult to puncture the sacroiliac joint. There are thick sacroiliac ligaments to penetrate and the joint is aligned oblique to the sagittal plane. The needle should be directed from a posterior approach a few centimetres away from the mid-line and pointing outwards. Oblique fluoroscopy is useful in planning the degree of obliquity. Care should be taken to recognize the posterior joint line on AP fluoroscopy avoiding confusion with the more lateral and more prominent anterior joint.

Injection is performed to introduce a diagnostic or therapeutic anaesthetic block or to obtain a bone specimen when infection is suspected. In difficult cases CT guidance may be required.

REFERENCES

1 PFENNINGER JL (1991). Injections of joints and soft tissues: Part 2. Guidelines for specific joints. *American Family Physician* **44**: 1196–1202.

2 KEROLUS G, CLAYBURNE G, SCHUMACHER HG (1989). Is it mandatory to examine synovial fluids promptly after arthrocentesis? *Arthritis and Rheumatism* **32**(3): 271–8.

3 HILL SA, MACLARNON JC, NAG D (1990). Ultrasound guided aspiration for transient synovitis of the hip. *Journal of Bone and Joint Surgery* (British volume) **72B**: 852–3.

4 ZAWIN JK, HOFFER FA, RAND FF, TEELE RL (1993). Joint effusion in children with an irritable hip: US diagnosis and aspiration. *Radiology* **187**: 459–63.

5 WILSON DJ (1990). Imaging of the temporomandibular joint. *TMJ Disorders*. A Wolfe Medical Atlas. Edited by Bramley P.

CHAPTER 2

Medical synovectomy

Nagui S. T. Gendi and John Outhwaite

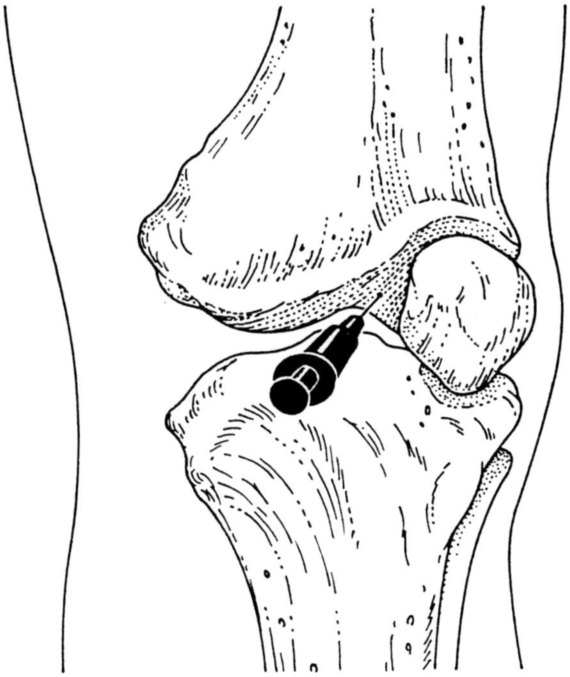

Medical synovectomy or synoviorthesis is the ablation of the synovial membrane of diseased joints through the intra-articular injection of chemicals or radioisotopes. Of the two forms, radiosynovectomy is the more commonly used. The use of intra-articular radiocolloids, using gold-198, was first reported by Ansell and colleagues[1] in 1963.

RADIOSYNOVECTOMY

In addition to gold-198, several other isotopes have been used in radiosynovectomy. These include yttrium-90,[2] dysprosium-165,[3] erbium-169,[4] phosphorus-32[5] and rhenium-186.[6] The choice is influenced by the characteristics of the individual isotope and the joint to be injected. The aim is to give local irradiation to the synovium with minimal leakage from the joint. Ideally, the radioisotope should be a pure beta (electron) emitter with no gamma photon emission. The latter contributes little to local irradiation, yet it adds to the total body dose, whereas beta particles typically penetrate around 1 cm of tissue and the effect is localized.

Spread is also influenced by the size of the radioactive particles, and efforts have been made to produce larger and more stable particles by using carriers such as ferric hydroxide macroaggregates[7] and polymeric microspheres.[8] Radioisotopes with shorter physical half-life have the advantage of undergoing rapid decay within the joint before significant leakage occurs. On the other hand, the half-life should be long enough to allow transport to the patient. The effective soft tissue penetration range should be equal to the thickness of synovium, which is in turn influenced by the size of the joint and the amount of synovial hypertrophy.

Yttrium-90 goes a long way in satisfying many of these requirements and is the most widely used isotope. Having the highest penetration range (maximum 10.8 mm, average 3.8 mm), yttrium-90 is particularly suited to treatment of the knee.[9] Although their half-lives are the same (2.7 days), yttrium-90 is preferred to gold-198 as it is a pure beta emitter and has a larger particle size.[10] The leakage of radioactivity to regional lymph nodes is less than 2 per cent for yttrium,[11] compared to more than 10 per cent in over a third of patients receiving gold-198.[12]

Phosphorus-32 has the disadvantage of a considerably longer half-life at 14.4 days[9] and its safety therefore remains to be established. Compared to yttrium-90, dysprosium-165 linked to ferric hydroxide macroaggregates has the attraction of a shorter half-life of 140 minutes, larger particle size and less leakage to regional lymph nodes at 0.17 per cent of injected dose.[13] However, the short half-life dictates that the reactor producing the isotope must be nearby. Its relative lack of penetration[9] (maximum 5.6 mm, average 1.3 mm) makes it less suitable for larger joints such as the knee. Agents with low penetration such as rhenium-186 and erbium-169 are recommended for medium-sized and small joints respectively.[4,6]

RHEUMATOID ARTHRITIS

The most common indication of radiosynovectomy is rheumatoid arthritis. Synovial inflammation is the hallmark of the disease and, if it persists, will lead to progressive joint damage. Efforts are directed at reducing the amount of joint inflammation by rest and the administration of non-steroidal anti-inflammatory agents, disease-modifying drugs, immunosuppressives and both systemic and local corticosteroids.

When synovial inflammation persists in one or more joints, surgical synovectomy can provide symptomatic relief for up to 5 years.[14] However, it should not be used lightly in the context of a chronic disease that often requires repeated surgical procedures each with its own risks and stress to the patient. In addition to the risks of open surgery and anaesthesia, surgical synovectomy is often followed by stiffness which will require prolonged in-patient physiotherapy.[15]

It is technically difficult to remove all parts of the synovium. The simple procedure of medical synovectomy offers an attractive alternative to surgery with comparable results.[16] The efficacy of radiosynovectomy has been confirmed by double-blind trials.[17,18] The results are better in those patients with minimal radiological changes: best in stage I with normal X-ray or osteoporosis, followed by stage II with slight cartilage or subchondral bone destruction, and less favourable or even poor in stages III and IV Steinbroker when there is prominent cartilage and bony destruction.[3,19,20]

As bone is less resistant than cartilage to radiological damage, radiosynovectomy should not be used where subchondral bone is exposed. It is less effective in joints with greatly hypertrophied synovium and multiloculated joints such as the wrist. It is not suitable for tenosynovitis. The effect is less when the rheumatoid disease is systemically active.[19,21]

The procedure can be repeated; the lack of initial response does not preclude a favourable response to a second injection.[21,22] As improvement may take time to become evident, the injection should not be repeated before 6 months has elapsed.[10]

Other joint diseases

Studies have frequently included cases of other inflammatory arthritides such as reactive arthritis, psoriatic arthritis and ankylosing spondylitis. It is likely that radiosynovectomy is as effective in these conditions as it is in rheumatoid arthritis.[23] Some have included patients with osteoarthritis, but the improvement appears to be short-lived.[20, 24] It has been shown[25] that radiosynovectomy is beneficial in pyrophosphate arthropathy. In pigmented villonodular synovitis, yttrium-90 has been successful in preventing recurrence after surgical synovectomy,[26] and also as an initial treatment.[27]

In haemophilic arthropathy, radiosynovectomy has been used to induce fibrosis of the subsynovial plexus. It prevents the recurrence of haemarthrosis in up to 60 per cent of cases, and reduces its frequency and severity in a further 28 per cent of cases.[28–30] Repeat injections can be effective in initially unresponsive joints. Radiosynovectomy is most effective in early arthropathy, and should be considered when more than three episodes of joint bleeding have occurred in 6 months. It has the advantage over surgical synovectomy in being much simpler, requiring much less rehabiliation and less factor VIII cover: 15 per cent for 48 hours as compared to 100 per cent reducing to 30 per cent over four weeks in the case of surgical synovectomy.[28]

Chromosomal abnormalities

Chromosomal abnormalities have been documented after radiosynovectomy.[31, 32] This is believed to be due to lymphocyte irradiation in regional lymph nodes since it is proportional to the amount of lymph node uptake and is reduced by immobilization.[33] The significance of these changes is unknown. They are transient and disappear within a few years of treatment.[28] There have been no reports of increased incidence of malignancy with radiosynovectomy during the 30 years since its introduction. Although it has been used in young patients, especially in the case of haemophilic arthropathy, it is best reserved for patients above the age of 45 when the child-bearing years have ended and when any theoretical risk of cancer induction is small compared to the naturally occurring risk.

Injection technique

The injection should be strictly intra-articular. In joints other than the knee, the position of the needle should be verified by arthrography, injecting a small amount of a contrast medium. Leakage of the radioisotope to the needle tract can cause radiation burns.[26] This can be prevented by avoiding overdistension of the joint through full joint aspiration prior to injection of the radioisotope and by flushing the needle and the needle track before the needle is withdrawn. A three-way stopcock can greatly facilitate the process of aspiration, injection and flushing.

Immobilization has been shown to reduce leakage of radioisotope to regional lymph nodes.[34] It is therefore important that the joint is immobilized using a splint or plaster of Paris. The duration of immobilization depends on the half-life of the isotope used. In the case of yttrium-90, it should be at least 48 hours. For dysprosium-165, 8 hours is sufficient. Premedication with intra-articular corticosteroids can also reduce leakage.[35] The same drugs are used to treat painful reactions to the injection.

The dose of radiopharmaceutical depends on the size of the joint and the radioisotope used. For the knee, 5 mCu (185 MBq) of yttrium-90 or 270 mCu (10 000 MBq) of dysprosium-165 is used. The dose of erbium-169 for the wrist, metacarpophalangeal and interphalangeal joints is 2 mCu (74 MBq), 1 mCu (37 MBq) and 0.5 mCu (18.5 MBq) respectively. The rhenium-186 dose is 2 mCu (74 MBq) for medium-sized joints. In the case of the knee, the technique of joint injection is as follows. The skin is sterilized using an alcoholic solution and the injection is performed using a non-touch technique. With the patient lying on the couch, the joint is punctured through medial or lateral approach at a point opposite the joint line, below the junction of the upper two-thirds and the lower third of the patella. The needle is directed horizontally under the patella and slightly cranially to avoid both the periosteum and the infrapatellar pad of fat (Fig. 2.1). The position of the needle is verified by aspirating synovial fluid. After full joint aspiration, the radioisotope, diluted in saline, is injected. Saline or lignocaine is used for flushing and is injected before and during needle withdrawal.

Chemical synovectomy

Chemical synovectomy is an alternative, especially for the younger patient in whom radiosynovectomy is contra-indicated. Thiotepa,[36] varicocid,[37] gold salts[38] and osmic acid[39] have been used. Osmic acid, which was introduced by Von Reis and Swensson in 1951,[40] is used most frequently, but its use should be limited to the knee joint.

The results of chemical synovectomy using osmic acid are comparable to those given by radiosynovectomy.[41, 42] As with radiosynovectomy, it is most effective in patients with little radiological changes.[39] It has been used in

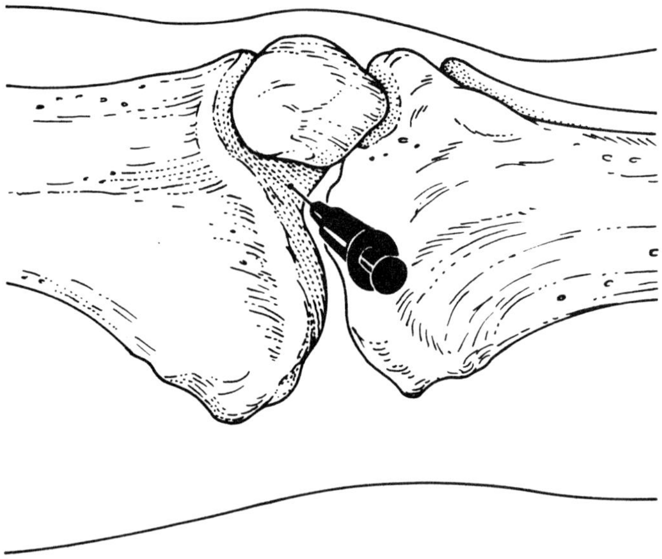

Fig. 2.1
Injection approach to the knee in medical synovectomy

children with juvenile chronic arthritis[43] and haemophilia.[44] Its value is in doubt, however, owing to concern about its long-term effect on the articular cartilage. Mitchell and colleagues,[45] using large doses of osmic acid in experiments on rat joints, demonstrated articular damage. When normal doses were used, in both rats and humans, the changes were mostly limited to the ultrastructural level and to the superficial layer of cartilage.[46,47] Moreover, clinical or radiological signs of accelerated cartilage damage have not been demonstrated.[41,43,48]

Side-effects are generally minor. Transient fever and leucocytosis occur in 15 per cent of patients. Local painful reactions are largely prevented by the concomitant use of lignocaine and corticosteroids. Some patients report transient urine discoloration. Fistulae and skin necrosis have been reported,[39,49] but can be prevented by careful technique.

The technique is similar to that of radiosynovectomy. Gloves and goggles should be worn for protection. The joint should be aspirated to dryness before injection to prevent overdistension and leakage into the needle track. In adults, 10 ml of 2% lignocaine is injected, followed by 10 ml of 1 per cent osmic acid and finally an injection of long-acting corticosteroid.

References

1 Ansell BM, Crook A, Mallard JR, Bywaters GLE (1963). Evaluation of intra-articular colloidal gold Au-198 in the treatment of persistent knee effusions. *Annals of the Rheumatic Diseases* **22**: 435–9.

2 Sheldon P, Webb C, Roy S, Nichol F, McKenzie J, Williams N, Zaphiropoulos G (1986). Treatment of chronic knee effusions with Yttrium-90: correlation of *in vitro* radiation and chemical effects with clinical results at three and six months. *Rheumatology International (Berlin)* **6**: 193–7.

3 Sledge CB, Zuckerman JD, Shortkroff S, Zalutsky MR, Venkatesan P, Snyder MA, Barrett WP (1987). Synovectomy of the rheumatoid knee using intra-articular injection of dysprosium-165–ferric hydroxide macroaggregates. *Journal of Bone and Joint Surgery* **69A**: 970–5.

4 Bouvier M, Bouysset M, Vonvoisin B, Diaine A, Lejeune E (1983). Synoviorthesès à l'erbium-169 et infiltrations d'exacetonide de triamcinalone dans les arthrites metatarsophalangiennes des rhumatismes inflammatoires chroniques. *Revue du Rhumatisme et des Maladies Ostéo-Articulaires* **4**: 267–71.

5 Onetti C, Gutierrez E, Hilba E, Aguirre CR (1982). Synoviortheses with ^{32}P colloidal chromic phosphate in rheumatoid arthritis: Clinical, histopathologic and arthrographic changes. *Journal of Rheumatology* **9**: 229–38.

6 Gamp R (1983). Die Radiosynoviorthese im Handbereich. *Aktuelle Rheumatologie* **8**: 165.

7 Chinol M, Vallabhajosula S, Zuckerman JD, Goldsmith SJ (1990). *In vivo* stability of ferric hydroxide macroaggregates (FHMA): Is it a suitable carrier for radionuclides used in synovectomy? *International Journal of Radiation Applications and Instrumentation, Part B* **17**: 479–86.

8 Mumper RJ, Mills BJ, Ryo UY, Jay M (1992). Polymeric microspheres for radionuclide synovectomy containing neutron-activated holmium-166. *Journal of Nuclear Medicine (New York)* **33**: 398–402.

9 Johnson LS, Yanch JC (1991). Absorbed dose profiles for radionuclides of frequent use in radiation synovectomy. *Arthritis and Rheumatism* **34**: 1521–30.

10 Lee P (1982) The efficacy and safety of radiosynovectomy. *Journal of Rheumatology* **9**: 165–8.

11 Gumpel JM, Williams ED, Glass HI (1973). Use of yttrium-90 in persistent synovitis of the knee. I. Retention in the knee and spread in the body after injection. *Annals of Rheumatic Diseases* **32**: 223–7.

12 Virkkunen M, Krusius FE, Heiskanen T (1967). Experiences of intra-articular administration of radioactive gold. *Acta Rheumatologica Scandinavica* **13**: 81–91.

13 Zuckerman JD, Sledge CB, Shortkroff S, Venkatesan P (1987). Treatment of rheumatoid arthritis using radiopharmaceuticals. *International Journal of Radiation Applications and Instrumentation Part B* **14**, 211-18.

14 Laurin CA, Desmarchais J, Daziano L, Gariepy R, Derome A (1974). Long-term results of synovectomy of the knee in rheumatoid patients. *Journal of Bone and Joint Surgery* **56A**: 521–31.

15 Camp AV (1972). Synovectomy of the knee in rheumatoid arthritis: Range of movement and quadriceps lag after surgery. *Annals of the Rheumatic Diseases* **31**: 155–8.

16 Gumpel JM, Roles NC (1975). A controlled trial of intra-articular radiocolloids versus surgical synovectomy in persistent synovitis. Lancet i: 488–9.

17 Bridgman JF, Bruckner F, Eisen V, Ticker A, Bleehen NM (1973). Irradiation of the synovium in the treatment of rheumatoid arthritis. *Quarterly Journal of Medicine. New Series* **42**: 357–67.

18 Delbarre F, Le Go A, Menkes CJ, Aignan M (1974). Preuve par étude statistique 'en double aveugle', de l'effect thérapeutique d'un colloide chargé d'yttrium radioactif (Y90) dans l'arthrite rheumatoide du genou. Comptes Rendus de l'Académie des Sciences; D: *Sciences Naturelles (Paris)* **279**: 1051–4.

19 Boerbooms AMT, Buijs WCAM, Danen M, van de Putte LBA, Vandenbroucke JP (1985). Radiosynovectomy in chronic synovitis of the knee joint in patients with rheumatoid arthritis. *European Journal of Nuclear Medicine* **10**: 446–9.

20 Spooren PFMJ, Rasker JJ, Arens RPJH (1985). Synovectomy of the knee with ^{90}Y. *European Journal of Nuclear Medicine* **10**: 441–5.

21 Winfield J, Gumpel JM.(1979). An evaluation of repeat intra-articular injections of yttrium-90 colloids in persistent synovitis of the knee. *Annals of the Rheumatic Diseases* **38**: 145–7.

22 Vella M, Zuckerman JD, Shortkroff S, Venkatesan P, Sledge CB (1988). Repeat radiation synovectomy with dysprosium 165–ferric hydroxide macroaggregates in rheumatoid knees unresponsive to initial injection. *Arthritis and Rheumatism* **31**: 789–92.

23 Topp JR, Cross EG, Fam AG (1975). Treatment of persistent knee effusions with intra-articular radioactive gold. *Canadian Medical Association Journal* **112**: 1085–9.

24 Will R, Laing B, Edelman J, Lovegrove F, Surveyor I (1992). Comparison of two regimes in inflammatory and osteoarthropathies. *Annals of the Rheumatic Diseases* **51**: 262–5.

25 Doherty M, Dieppe PA (1981). Effect of intra-articular yttrium-90 on chronic pyrophosphate arthropathy of the knee. *Lancet* ii: 1243–6.

26 Gumpel JM (1986). Intra-articular therapy. In: *Copeman's Textbook of the Rheumatic Diseases*, pp. 587–91. Edited by Scott JT. Churchill Livingstone, Edinburgh.

27 O'Sullivan MM, Yates DB, Pritchard MH (1987). Yttrium synovectomy: a new treatment for pigmented villonodular synovitis [letter]. *British Journal of Rheumatology* **26**: 71–2.

28 Fernandez-Palazzi F, de Bosch NB, de Vargas AF (1984). Radioactive synovectomy in haemophilic haemarthrosis: Follow-up of fifty cases. *Scandinavian Journal of Haematology* **33** (Suppl. 40): 291–300.

29 Ortonowski G, Ziemski JM, Kusharski W, Woy-Wojciechowski J (1990). Synoviorthesis with ^{198}Au colloid gold in haemophilia patients. A preliminary report. *Folia Haematologica (Leipzig)* **117**: 505–10.

30 Rivard GE, Girard M, Cliche CL, Guay JP, Belanger R, Besner R (1982). Synoviorthesis in patients with haemophilia and inhibitors. *Canadian Medical Association Journal* **127**: 41–2.

31 Doyle DV, Glass JS, Gow PG, Daker M, Grahame R (1977). A clinical and prospective chromosomal study of yttrium-90 synovectomy. *Rheumatology and Rehabilitation* **16**: 217–22.

32 Stevenson AC, Bedford J, Dolphin CW, Purrot RJ, Lloyd DC, Hill AGS, Hill HFH, Gumpel JM, Williams D, Scott JT, Ramsey HW, Bruckner FE, Fearn BDA (1973). Cytogenetic and scoring study of patients receiving intra-articular injections of gold-198 and yttrium-90. *Annals of the Rheumatic Diseases* **32**: 112–22.

33 Gumpel JM, Stevenson AC (1975). Chromosomal damage after intra-articular injections of different colloids of yttrium-90. *Rheumatology and Rehabiliation* **14**: 7–12.

34 Roberts SD, Gillespie PJ (1973). Effect of immobilization on retention of ^{90}Y. *Annals of the Rheumatic Diseases (supplment)* **32**: 46–7.

35 Goode JD, Howey S (1973). Effect of premedication with intraarticular hydrocortisone on the retention of Au198 in the knee joint in rheumatoid artritis. *Annals of the Rheumatic Diseases (Suppl.)* **32**: 43–5.

36 Ellison MR, Flatt AE (1971). Intra-articular thiotepa in rheumatoid disease: A clinical analysis of 123 injected MP and PIP joints. *Arthritis and Rheumatism* **14**: 212–22.

37 Hafner R, Truckenbrodt H (1985). Die Synoviorthese mit Varicocid in der Behandlung der juvenilen chronischen Arthritis. *Aktuelle Rheumatologie* **10**: 202.

38 Lewis DC, Ziff M (1966). Intra-articular administration of gold salts. *Arthritis and Rheumatism* **9**: 682–92.

39 Sheppeard H, Ward DJ (1980). Intra-articular osmic acid in rheumatoid arthritis: five years' experience. *Rheumatology and Rehabiliation* **19**: 25–9.

40 Von Reis G, Swensson Å (1951). Intra-articular injections of osmic acid in painful joint afflictions. *Acta Medical Scandinavica Supplement* **259**: 27.

41 Sheppeard H, Aldin A, Ward DJ (1981). Osmic acid versus yttrium-90 in rheumatoid synovitis of the knee. *Scandinavian Journal of Rheumatology* **10**: 234–6.

42 Nissila M, Anttila P, Hamalainen M, Jalava S (1978). Comparison of chemical, radiation and surgical synovectomy for knee joint synovitis. *Scandinavian Journal of Rheumatology* **7**: 225–8.

43 Martio J, Isomaki H, Heikkola T, Laine V (1972). The effect of intra-articular osmic acid in juvenile rheumatoid arthritis. *Scandinavian Journal of Rheumatology* **1**: 5–8.

44 Menkes CJ, Allain JP, Genil C, Witvoet H, Tak-Tak H, Simon F, Delbarre F (1973). La synoviorthèse à l'acide osmique chez l'hémophilie. *Revue du Rhumatisme et des Maladies Osteo-Articulaires (Paris)* **40**: 255–8.

45 Mitchell N, Laurin C, Shepard N (1973). The effect of osmium tetroxide and nitrogen mustard on normal articular cartilage. *Journal of Bone and Joint Surgery* **55B**: 814–21.

46 Niculescu D, Tomescu E, Ionescu C, Cotutiu C, Hrisanidi St, Negoescu M, Cindea C, Stoia I (1976). Ultrastructural changes of cartilage after intra-articular administration of osmium tetroxide and the sodium salts of fish oil fatty acids. *Scandinavian Journal of Rheumatology* **5**: 133–40.

47 Nissila M, Ahlqvist J, Collan Y, Raunio P, Isomaki H (1977). Morphological findings in joint cartilage after osmic acid treatment. *Scandinavian Journal of Rheumatology* **6**: 231–6.

48 Nissila, M.(1978). Absence of increased frequency of degenerative joint changes after osmic acid injections. *Scandinavian Journal of Rheumatology* **7**, 81–4.

49 Jean P, Jullian E, Rodor F, Kevorkian B, Jean-Pastor MJ, Gailhaguet AM, Jouglard J (1986). Nécrose cutanée consécutive à une synoviorthése au tétroxyde d'osmium. *Therapie* **41**: 357–9.

2

MASS LESIONS

CHAPTER 3

Abscess drainage

David J. Wilson

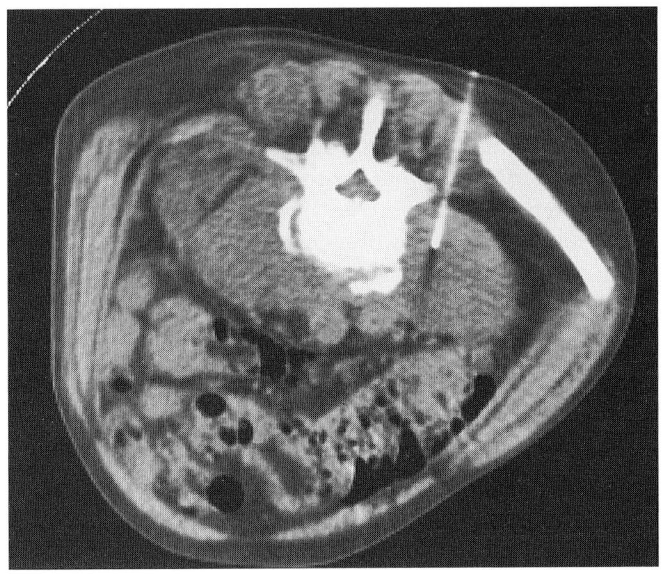

Introduction

An abscess is a collection of pus that is discrete from the surrounding tissues. The majority of abscesses are due to infection which develops from blood-borne organisms, by local spread of sepsis or by implantation of organisms. A pseudo-capsule consisting of debris, necrotic tissue and inflammatory reaction typically forms a wall or capsule around the infected cavity. Almost any part of the body may be affected and the symptoms will depend on the normal function of the areas involved. Osteomyelitis may present as an erupting soft tissue abscess communicating with the bone infection. Potential spaces between organs and in fascial planes are common sites for abscesses. They vary in size from a few millimetres in diameter to huge collections containing several litres of pus (Fig. 3.1).

Normally the infecting agent may be isolated from the pus and its sensitivity to antibiotics is then determined. In all but the smallest of lesions antibotics are unlikely to penetrate to the centre of the cavity owing to poor or absent blood supply and anatomical disruption. For this reason, when there is a true abscess cavity, drug therapy alone is unlikely to be sufficient and drainage is essential.

Diagnosis

Clinical

Pyrexia and systemic toxicity are typical features. General debility and localized pain are common. The affected part is immobile and disabled. Sometimes a mass is present which may point to the skin surface and eventually discharge and drain. The route of extension and drainage is usually thorough tissue planes that offer the least resistance, although this may be a convoluted route with distant drainage, a fact that may cause confusion as to the origin of the pus. Patients who are immunosuppressed from whatever cause may exhibit few of these signs and symptoms, and the attending clinician must have a high index of suspicion for deep infection.

Laboratory investigations commonly show signs of inflammation with a raised ESR, abnormal white cell count and anaemia. Again systemic disease and immune compromise may mask these signs. In addition, the interpretation of results may be difficult and normal findings cannot be used to exclude deep sepsis.

Imaging

Imaging studies are useful for the detection or exclusion of focal sepsis and for the staging of the extent of the lesion. They are valuable in guiding percutaneous therapy and monitoring the progress of the disease.

DETECTION

If there are local symptoms or signs they will serve as the initial guide to investigation. The most appropriate technique will depend on the structures involved. The choice of method follows the same guidelines described below for abscess staging.

In the absence of localizing features the technique must search large areas of tissue seeking focal change in anatomy or physiology. Plain radiographs do not adequately show soft tissues. Ultrasound is a technique best suited to detailed examination of small areas and not for survey of large areas.[1] CT, although effective in the detection of deep infection, has a significant penalty in radiation dose. Nuclear medicine and MRI are the methods best suited to screening for deep or occult infection.

Nuclear medicine

Large-field gamma cameras allow a survey of the whole body, a technique that is especially suited to the detection of metastatic malignancy but one that also works well for occult infection. There are a variety of suitable radiopharmaceuticals. Products labelled with technetium-99^{m} are optimum for imaging, as the short half-life (6 hours) keeps radiation dose low. The isotope is a pure gamma emitter and it is readily produced from a molybdenum-99 generator, which has a reasonable shelf life (half-life 67 hours). There are minimal radioactive contaminants and the gamma energy (140 kV) is particularly suitable for detection by the sodium iodide crystal within a gamma camera. Physiologically, technetium resembles potassium and is not known to have any deleterious effect on the body when injected.

Linked to pharmaceutical agents technetium-99^{m} can be used a tracer of physiological and pathological activity. Hydroxy-mono-diphosphonate (HMDP) is a bone-seeking agent which is commonly used to seek metastatic disease but also has a role in detecting both bone and, perhaps surprisingly, soft tissue infection. Areas of increased bone metabolism will be active on ^{99m}Tc HMDP studies whilst dead bone will give a less intense image than normal bone.

Uptake depends on several factors. Around 60 per cent equilibrates with bone whilst 40 per cent is excreted via the kidneys. Cardiac output and renal function affect general distribution and if severely impaired may prevent adequate examination. Radiopharmaceutical will collect in areas of increased osteoid formation, areas of higher blood flow, areas of increased osteoid mineralization and where the sympathetic nerve supply has been impaired.

Infection produces a variety of pathological processes including areas of inflammation, areas of repair and areas

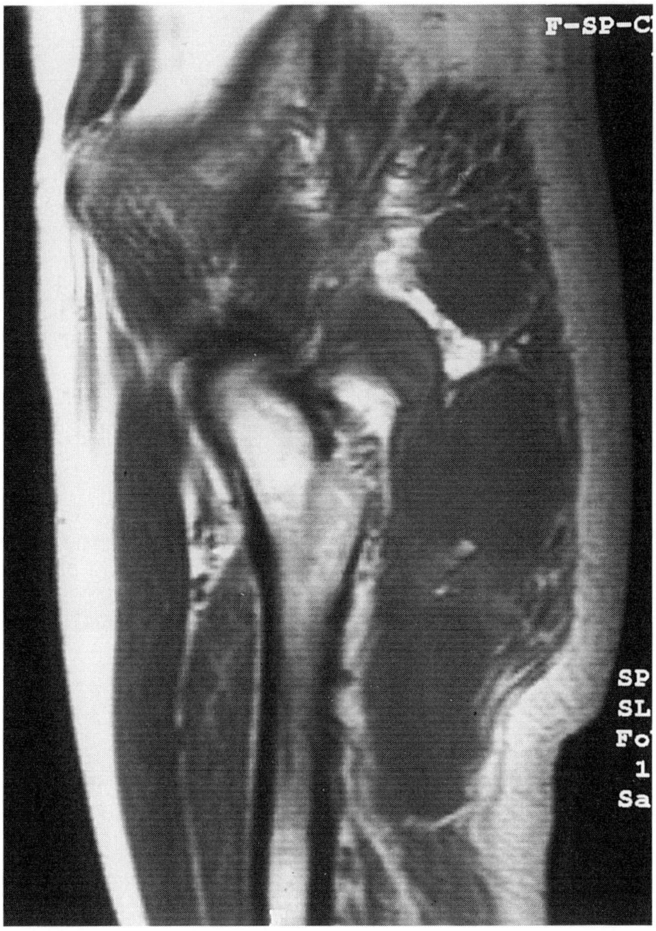

(a)

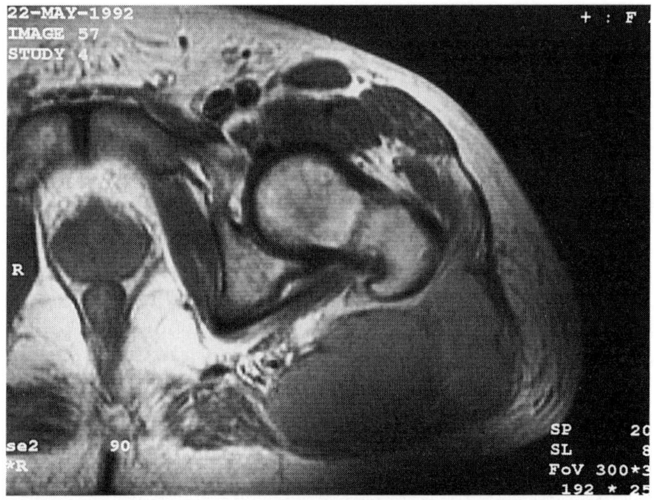

(b)

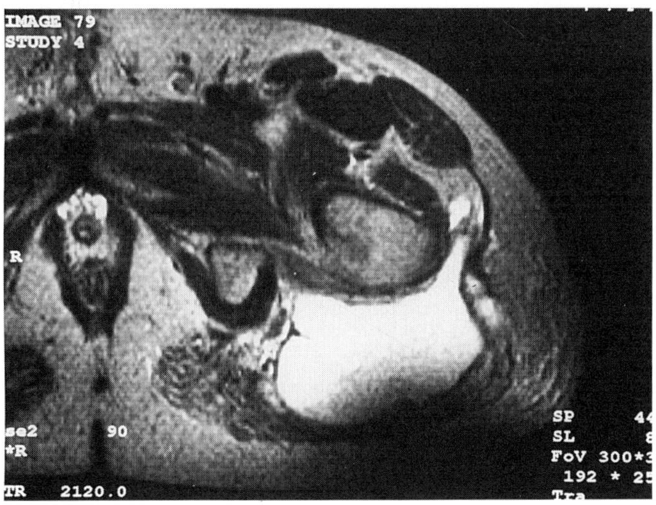

(c)

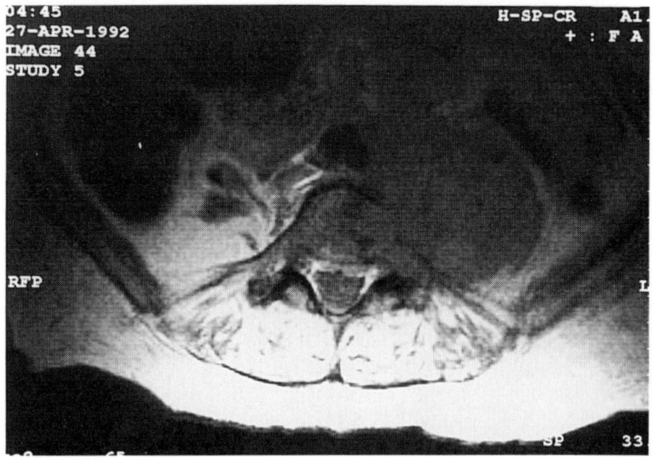

(d)

Fig. 3.1
Magnetic resonance imaging of a massive abscess extending from the sciatic region to the thigh. Higher sections demonstrated that this arose from a psoas and retroperitoneal collection and probably originated as a disc space infection. (a) Sagital T1 weighted spin echo: a multi-loculated collection with walls of different signal. The adjacent structures are displaced rather than invaded. (b) Axial T1 weighted spin echo: images in several planes clarify the anatomy. (c) Axial T2 weighted fast spin echo: T2 weighted images show fluid, pus and for that matter solid tumour as intense white (high signal). The anatomical resolution is less precise but the discrimination of abnormal tissue is much improved as compared with T1 weighted images. (d) Axial T1 weighted spin echo at L5 (some motion artefact). The abscess was followed upward through a thin track that extended through the sciatic notch to a mass in the left iliopsoas muscle. The adjacent vertebral outline was impaired. A disc space origin seems likely.

of ischaemia and bone death. Therefore infection is variously associated with increased blood supply (increased activity on blood pool images), decreased blood supply (cold areas on blood pool and delayed images) and increased bone turnover (increased activity on crystal phase images). As a result, infection in bone may be detected with great sensitivity. Fortunately, abscess cavities in soft tissues and diffuse cellulitis also show increased activity with bone-seeking agents. Blood pool or dynamic images show increased uptake in soft tissues due to hyperactive circulation. Dead or infarcted tissue may take up this radiopharmaceutical. It should be remembered that these studies are sensitive but non-specific. Most disease states, including tumours and

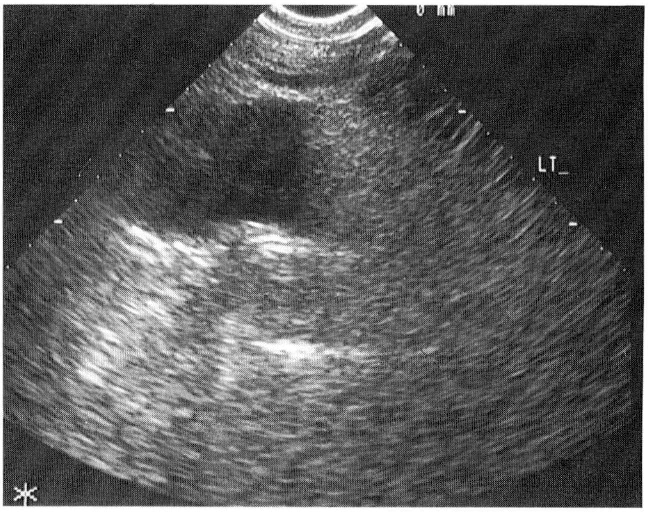

(a)

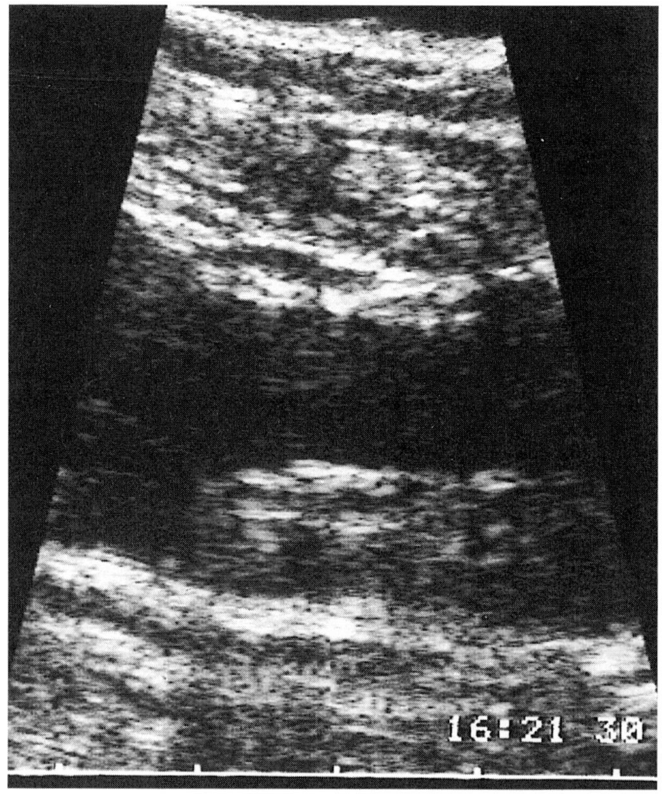

(b)

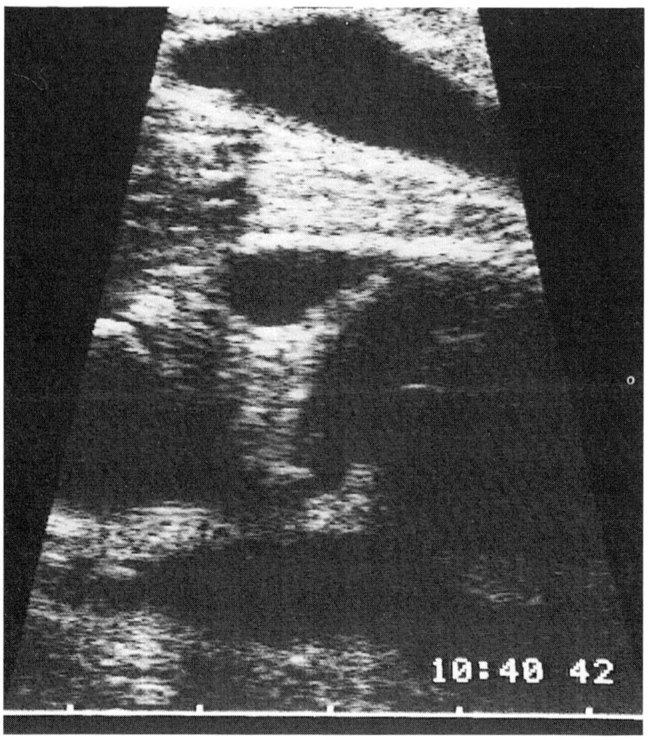

(c)

Fig. 3.2 *Ultrasound of abscesses*
(a) Fluid is echo-free (black) on ultrasound. Pus may contain low-level echoes and floating, more solid material. Acoustic enhancement is seen behind the dark area in this abscess, seen as a bright shadow behind the collection.
(b) Solid material within a fluid collection. (c) A loculated collection.

trauma, give similar appearances. Other imaging may help to differentiate but microbiological and histological studies are usually indicated to determine the cause of the abnormal activity.

Indium-labelled white cells, labelled with the radiopharmaceutical *in vitro* or *in vivo*, have been advocated as a more specific means of detecting infection. Similarly gallium citrate has been used as a means of increasing specificity. Although targeted to soft tissue disease, these agents are, unfortunately, less sensitive than ^{99m}Tc products. Gallium-67 produces radiation that is more penetrating and the collimation is less precise. Indium labelling is a complex and time-consuming procedure and the radioisotope is often contaminated with other non-contributory isotopes, increasing the radiation dose received by the patient. Neither technique is ideal, although there will be a few instances where occult infection is detected by these selective methods when ^{99m}Tc bone-seeking agents are unhelpful. With the advent of MRI and percutaneous biopsy it is doubtful whether there is any residual role for the more selective nuclear medicine studies in defining the cause of abnormal bone scintigraphy or localized symptoms.

Nuclear medicine studies are enhanced in sensitivity and precision by tomographic imaging (emission computerized axial tomography, ECAT) but they are still relatively indistinct compared with other cross-sectional methods and are not usually used to guide therapeutic manoeuvres.

Magnetic resonance imaging

Recently, MRI using a wide field of view has been advocated for the detection of occult sepsis. Dead tissue will show low or absent signal on all sequences. Pus will appear as fluid with low signal on T1 weighted images and high signal on T2 and T2* weighted images. Note that gradient echo imaging produces susceptibility artefact in trabecular bone, masking any effect of disease on the fatty marrow. These sequences are insensitive to marrow disease. Similarly, fast spin echo T1 weighted images of marrow show less contrast between the low-signal areas of marrow replacement and the residual marrow than is seen on conventional T1 weighted spin echo studies. They save only a little time and should not be used when sepsis is suspected. Oedema and fluid show low signal on T1 weighted images and high signal on T2 weighted images (Fig 3.1c). Fat suppression by radio frequency presaturation will diminish the high fat signal making areas of oedema more obvious. Inhomogeneity in the radio frequency suppression may give artefacts and lead to confusing images.

A better means of removing the fat signal is to rely on short Tau inversion recovery (STIR) sequences. These are heavily T2 weighted but show little signal from fat. Until recently the long imaging time and poor signal-to-noise ratio detracted from their value but now the application of fast spin echo methods to the STIR sequence allows excellent-quality images of large field of view (50 cm plus) in times under 5 minutes. Although further studies are required to determine the sensitivity of these methods compared to nuclear medicine studies, subjective early work looks encouraging.

LOCAL STAGING

Plain radiographs aided by tomography and CT will show bone involvement. Soft tissue changes may be studied by ultrasound,[2,3] CT or MRI to define the location and spread of the disease.

Ultrasound

Readily available and reasonable cheap to perform, examination using ultrasound lends itself to the detection of abscesses. Fluid is echo free and pus usually has a low level of echoes compared to muscle and fat. Acoustic enhancement occurs beyond or below fluid, casting a bright shadow. This is a technical artefact produced because the sound travelling through the fluid is less attenuated than the sound travelling through adjacent solid material, leading to a brighter echo. The same effect occurs with solid material that has a homogeneous texture with few internal echoes. Hence some solid mass lesions may be confused with fluid collections. Lymphomas and neuromas are especially prone to this effect. To avoid this pitfall the examiner should place the transducer over an area of known fluid at the same depth; blood vessels or the bladder are convenient. Adjusting the gain settings to the point at which fluid is just showing a few internal echoes and then returning to the suspect areas will clinch the diagnosis, as solid homogeneous lesions will have many more internal echoes at these settings.[4]

Soft tissue collections are readily seen provided that they are not masked by gas in lung or bowel and that they do not lie behind bone (Figs 3.2 and 3.3). Bone reflects all sound and creates some reverberation artefact that may show as lines across that adjacent soft tissue. Gas also blocks sound whether in bowel or lung. Normal muscle shows a 'herring bone' pattern that is disrupted by cellulitis and replaced by fluid in abscess formation. Each muscle gives a different echo pattern and before diagnosing an abnormality the observer should examine the opposite, normal, side. In practice these features mean that ultrasound is excellent for screening for suspected soft tissue infection but the psoas and paravertebral areas are difficult to visualize. Osteomyelitis commonly is associated with periosteal collections of fluid which may be seen on ultrasound. A normal study does not exclude disease in these deep locations but, even so, ultrasound is often diagnostic.[5] The method requires experienced observers and it is very difficult to produce images that may be interpreted away from the patient. Follow-up studies must be planned carefully to obtain equivalent sections and are best performed by the same observer.

Computed tomography

By comparison CT is easy to repeat and is particularly suitable for following the progress of an infected area (Fig. 3.4). Unfortunately the fairly high radiation dose prohibits its liberal use, and repeat studies must be limited in extent and frequency. Soft tissue infection changes the architecture of the muscles, produces mass effects and leads to increased X-ray attenuation in adjacent fat (the dirty fat phenomenon). The oedematous and swollen tissue may have attenuation similar to that of normal muscle, and intravenous contrast enhancement is often required to discriminate the margins of the lesion. Bone infection changes the trabecular pattern and may be associated with periosteal new bone and soft tissue extension.

Magnetic resonance imaging

To date MRI is the optimum method for localizing and staging infection (Figs 3.1 and 3.5). The techniques already described allow detailed anatomical examination and clear discrimination of abnormal from normal tissue. However, the tissue specificity of MRI should not be overestimated as oedema looks like infected tissue and malignant invasion. Some degree of tissue characterization

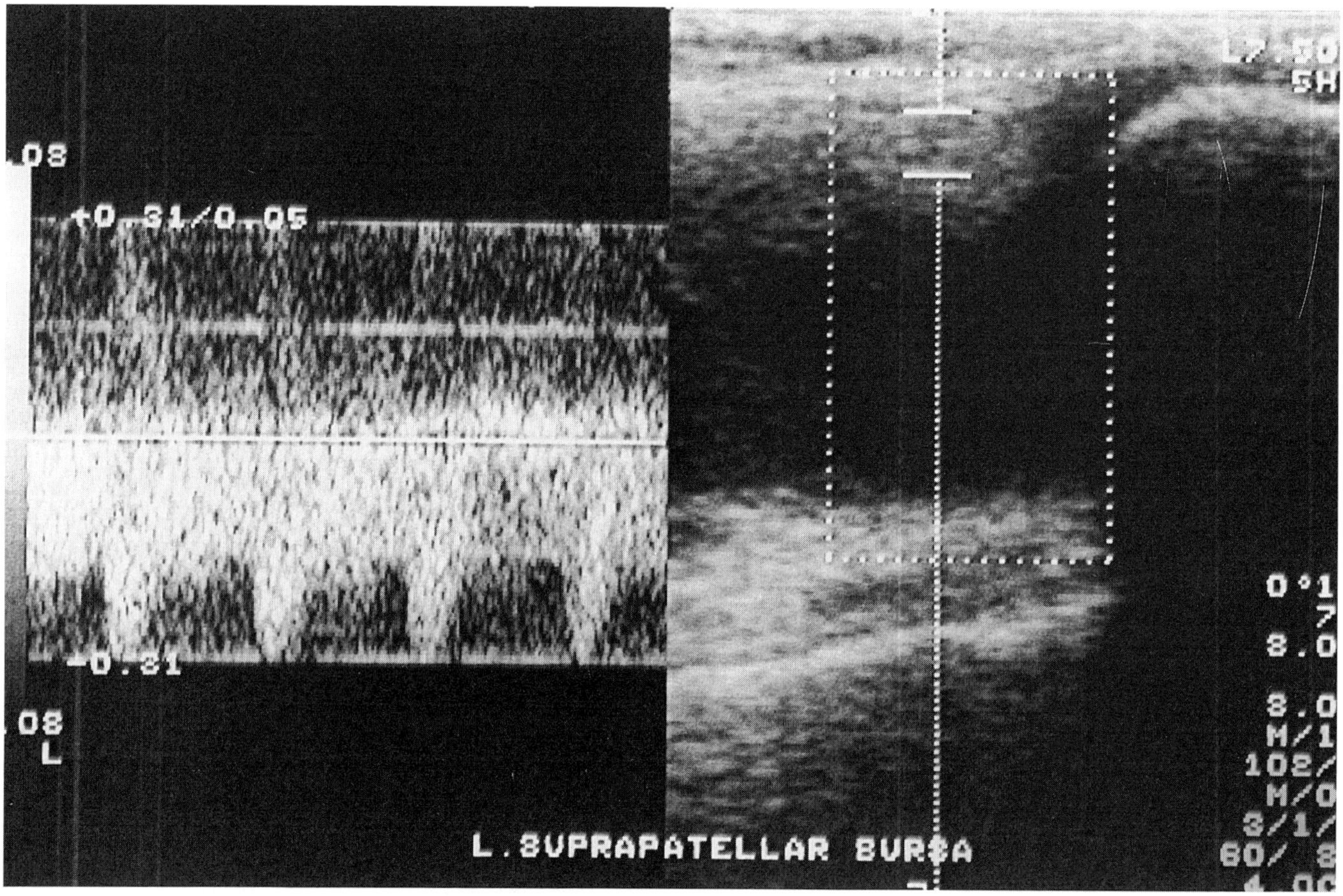

Fig. 3.3
Septic arthritis of the knee with a large suprapatellar collection. The Doppler study shows flow within the thickened synovium.

may be achieved as fat and altered blood have typical appearances. MRI may be safely repeated and its availability is increasing whilst costs are falling.

GUIDANCE

In selecting the optimum method to guide needle placement the primary consideration is which technique best demonstrates the abnormality. In the case of musculoskeletal sepsis this may be fluoroscopy for bone infection, ultrasound for soft tissue infection and CT or MRI for both. Many ultrasound machines have biopsy attachments with electronic guide tracts on screen. These allow precise guidance but they are clumsy to use and create problems in maintaining sterility. Many prefer to locate the collection, measure the depth and observe its location, then place the needle without direct guidance. Alternatively they may place the probe oblique to the needle track away from the sterile area and then observe the needle tip by the bright echo that it commonly produces. Some needles are designed to give a prominent echo from the tip. An alternative is to inject local anaesthetic or saline. The microscopic bubbles in the fluid give bright moving echoes close to the needle tip.

When using fluoroscopy every effort should be made to minimize the radiation dose to the patient and the operator. Lead protection and tight coning of the beam are essential. The hands of the operator should never enter the beam even if enclosed in lead gloves as the gloves are not designed to protect against exposure to the main beam. The procedure should be: look at the TV monitor, switch on, freeze the image and switch off, decide on the next move, smoothly move the needle about 1–2 cm with the fluoroscopy switched off, and repeat. Small, smooth movements are much more comfortable for the patient and less likely to move irremediably off the desired track.

Placement of the needle or catheter will require many readjustments of position, and a dynamic or real-time technique makes these much easier. Therefore ultrasound or fluoroscopy have great advantages. CT-guided biopsy or drainage is a relatively lengthy procedure and depends on accurate initial needle placement. Metal artefact may confuse the images and if the needle slips away from the exact plane of the slice it may appear to be inserted less deeply than it really is. Measurement of the

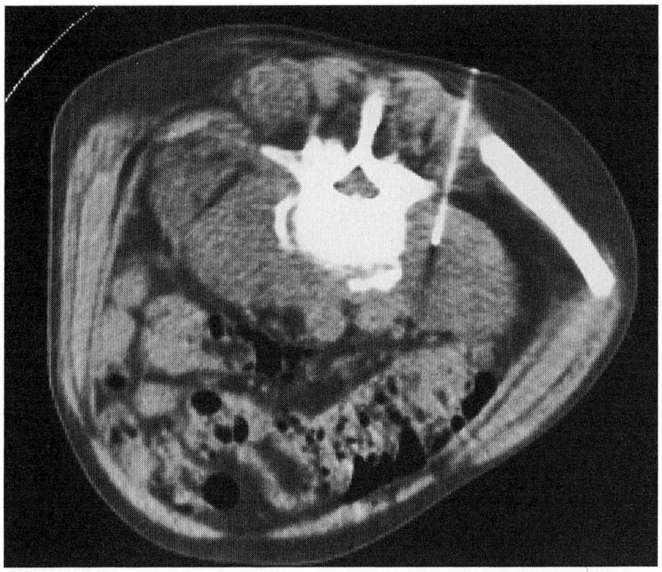

(a)

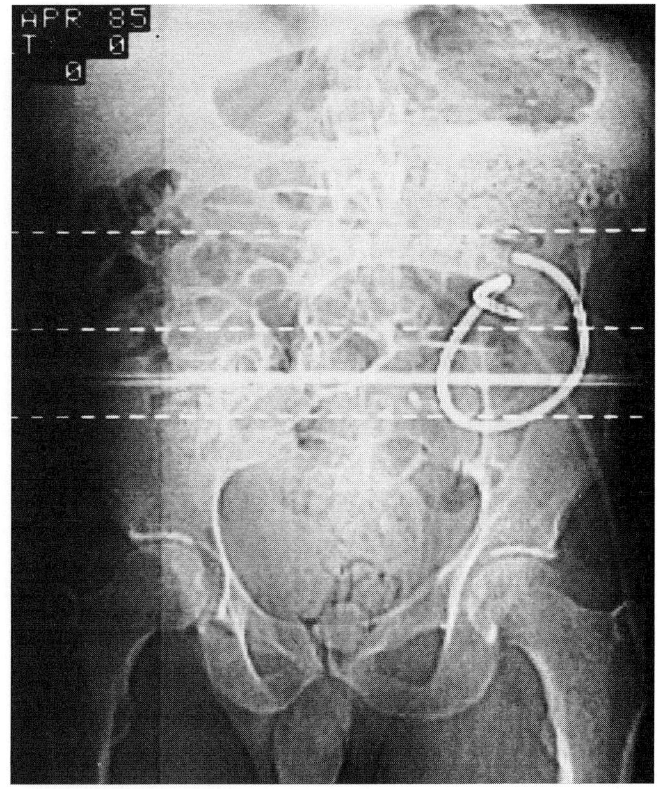

(b)

Fig. 3.4
(a) CT guided aspiration of a psoas abscess. (b) The scout view shows the position of the drainage catheter inserted over a guide wire. A second scout taken in the lateral projection will suffice as a post-procedure plain film examination and save moving a sedated patient to another room. With permission of *Current Orthopaedics*, Churchill Livingstone.

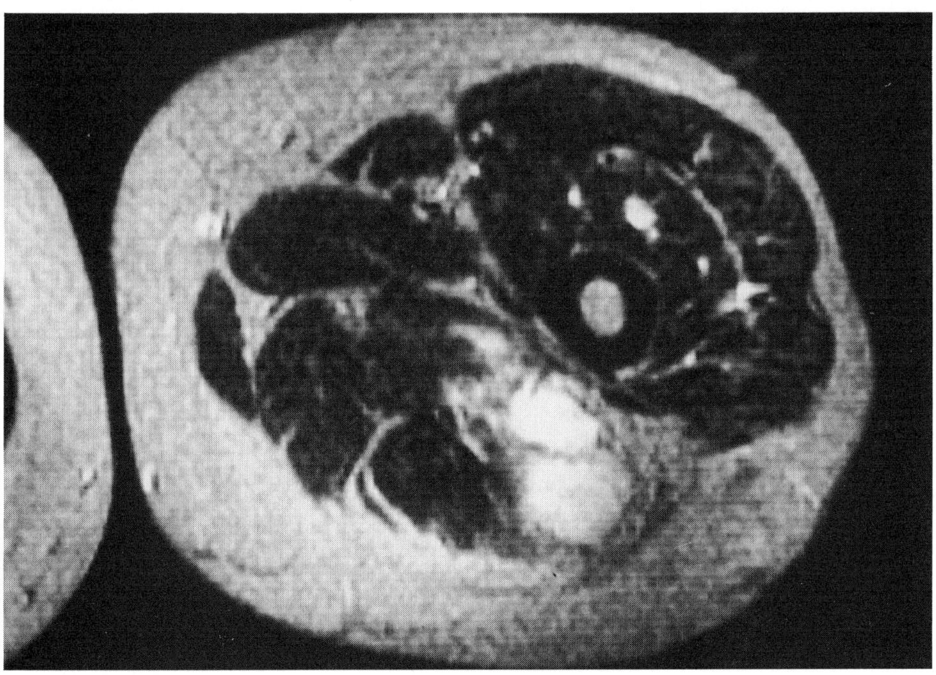

Fig. 3.5
A small abscess in the posterior compartment of the thigh seen clearly on T2 weighted fast spin echo MRI.

depth of the lesion and the length of inserted needle helps to avoid this pitfall. In addition, the needle tip may exhibit a low attenuation artefact. Some forethought is vital as the gantry aperture may be too small to allow the patient to enter with the needle inserted. MRI has all these drawbacks as a guidance technique plus the problems of access to a patient deep in the tunnel and the sometimes huge artefact produced by the needle. Of course it is mandatory to use MRI-safe needles, which are inevitably more expensive.

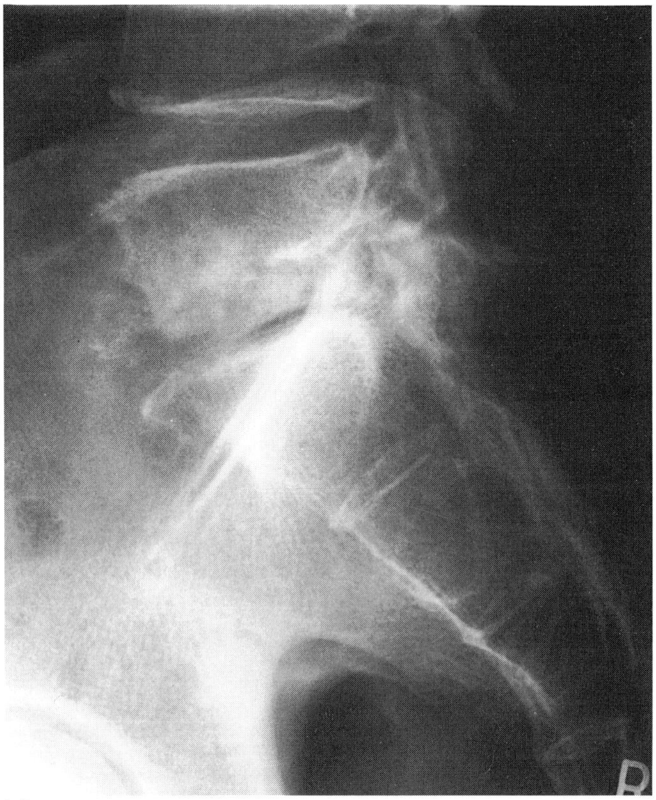

(a)

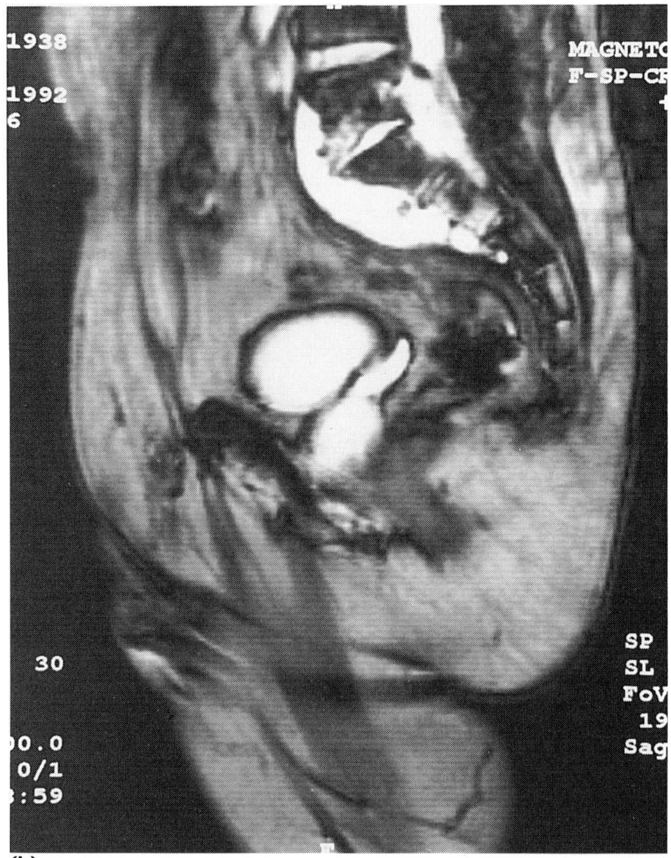

(b)

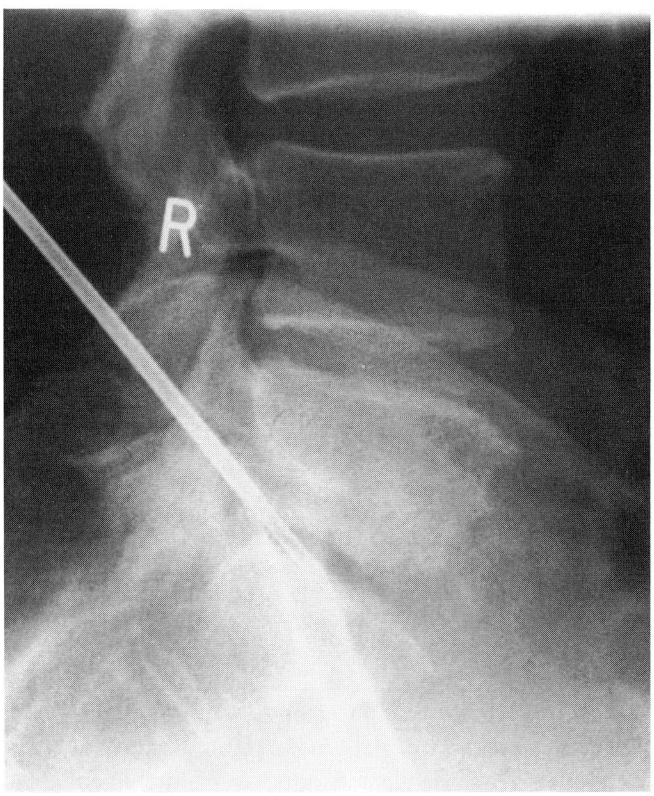

(c)

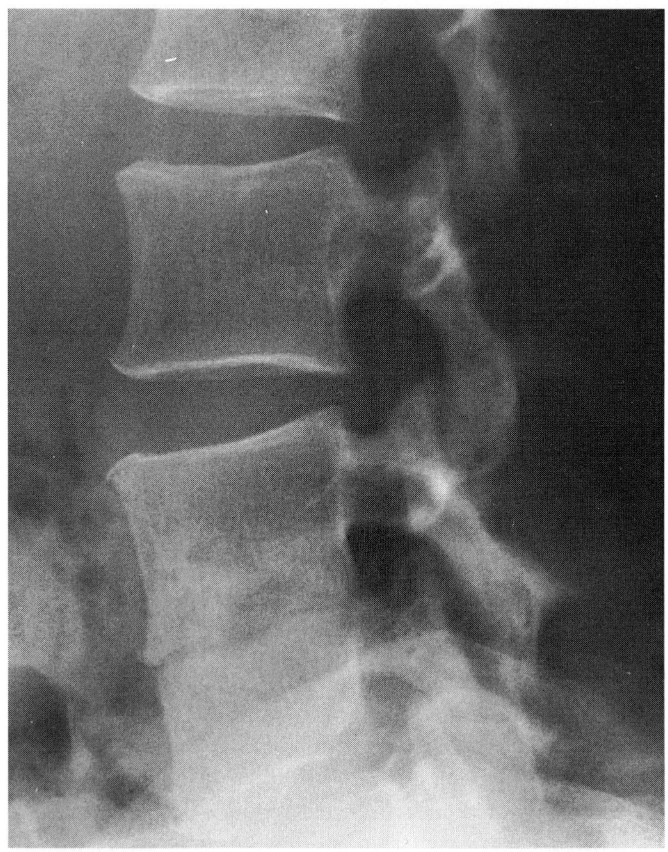
(d)

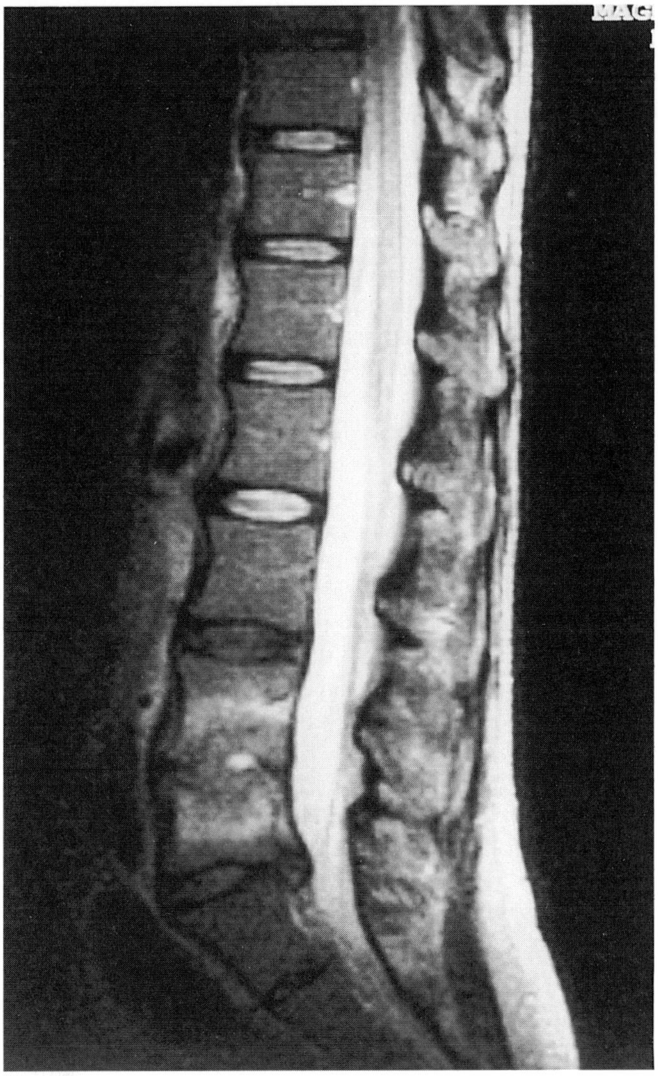

(e)

Fig. 3.6
Spinal infection commonly arises in disc spaces. Plain films are usually diagnostic but MRI is required to show the extent of soft tissue involvement and any effects on the spinal cord or nerve roots. Biopsy is essential to determine the organism and should include a bone specimen as the histological evidence is frequently important. (a) Plain radiograph showing a narrowed L5/S1 interspace with adjacent bony destruction anteriorly. Note that the vacuum effect is confined to the posterior interspace. Loss of a vacuum effect on serial films is circumstantial evidence of sepsis. (b) Sagittal body coil T2* weighted gradient echo MRI shows a surrounding abscess and bone oedema (bright signal areas). On this sequence disc material normally has a high signal. (c) Trephine percutaneous biopsy confirmed the organism. Note that the needle is directed through this disc space into bone. (d) Plain film in a second case showed disc space narrowing but no frank bone destruction. (e) Sagittal T2 weighted fast spin echo MRI shows adjacent bone oedema but no no soft tissue extension. The differential diagnosis was of stress change in a degenerate disc disease (osteochondrosis), but loss of the endplates was suspicious. Bone biopsy confirmed infection.

In practice ultrasound and fluoroscopy are the preferred methods, reserving CT guidance for those deep structures where precise depth and approach are essential to avoid structures that are not well seen with ultrasound or fluoroscopy. For example, bowel gas may obscure retroperitoneal structure in anterior approach through the pelvis, and lung often obscures paravertebral collections. In these circumstances CT guidance is most effective. MRI guidance is indicated only when it is the only means of seeing the lesion. Even then it is often better to identify adjacent landmarks that will be visible on X-ray from the MR study and use fluoroscopy for the needle guidance.

Diagnostic aspiration

Investigation of suspected soft tissue or bone infection must include bacteriological and, often, histological examination of tissue. Cultures from a discharging sinus are easy to take but commonly fail to isolate the causative organism. Secondary sinus infection is frequent and the cultures will be unreliable. The best results will be achieved by needle aspiration or biopsy. If bone is involved it is prudent to include a trephine specimen for histological examination (Fig. 3.6). The choice of guidance technique depends on the site of the lesion, the surrounding anatomy and the practical drawbacks of each technique as discussed above. The selection of the needle depends on the location and likely depth. As pus may be thick and glutinous, a wide-bore needle should be employed; normally around 18 G is suitable. The needle should have a trocar which is not removed until it is fully in place. This prevents plugging with tissue prior to aspiration. As infection may be confused with neoplasia and vice versa it is prudent to perform both bacteriological and histological studies on all specimens. To obtain tissue a side cutting (Trucut style) needle is ideal, end cutting (Menghini style) needles rarely obtain adequate tissue from necrotic material. If, bone is involved a trephine needle with an internal diameter of at least 2 mm should be used. The architecture of the trabeculae is important in histological diagnosis, and is difficult to study in small fragments.

The patient should be prepared with sedation and analgesic premedication selected by judging each case individually. Adults rarely require a general anaesthetic but children are a different matter. Babies have no anxiety in anticipation but they do feel pain and the dose of analgesia and sedatives is critical. They must be prescribed and monitored by an experienced clinician or anaesthetist. Infants and children up to the early teens often require heavy sedation or general anaesthesia to achieve the necessary immobility. In general it is wise to rely upon a paediatric anaesthetist to achieve the required level of co-operation from the child. If heavy sedation is used in adults it is wise to have a medically qualified assistant to monitor the patient as such a person will not be distracted by the complexity of the biopsy. Appropriate resuscitation equipment and staff should be to hand with reversant drugs for the sedatives employed. Local anaesthetic is introduced for all but the simplest of punctures and careful aseptic technique is used both to protect the patient and to prevent contamination of the specimen.

The route of approach should be planned to avoid important structures whilst considering that a surgeon may need to resect the track if the diagnosis of infection is incorrect and the lesion proves malignant. For moderate-sized abscesses and large lesions the needle tract may become the route of pointing and discharge. The route should also be planned bearing in mind the nursing problems that may result. For example, it is better to approach a psoas lesion from the flank to make nursing of the drainage route easier but it should not be so far lateral that the bowel is traversed or the peritoneal cavity penetrated.

Diagnostic aspiration may be part of the therapeutic procedure when the case is clear enough to plan both in advance.

Specimens

There is strong evidence that organisms from musculoskeletal infection may be sparse and difficult to isolate. Local anaesthetic is itself strongly antibacterial.[6] The microbiologist performing cultures and analysis should be consulted regarding the handling of the specimen. Often they will ask for inoculation of material into culture-enhancing media immediately after its aspiration. Delicate anaerobic organisms may survive only if cultured from solid tissue biopsy. In general the larger the piece of tissue the better. Similarly, advice should be sought in advance as to how to handle histological material. Placing specimens into formal saline may delay decalcification of bone-containing fragments and the laboratory may prefer fresh and untreated material delivered without delay or desiccation. If bone is involved then it is common for the cultures to be negative, whilst the histological appearances of the bone biopsy or, less often, the adjacent soft tissues show changes that are diagnostic of sepsis. This will encourage a more rigorous search for the causative organism or, if there are caseating granulomata, will clinch the diagnosis.

Therapeutic aspiration

Abscess cavities are surrounded by tissue with relatively active blood supply but it is common for antibiotic agents to fail to reach therapeutic levels at the centre of the lesion. The debris and necrotic material will create a painful mass and lead to continued general morbidity. Abscesses should be drained as much as practical by the initial diagnostic aspiration but they will reaccumulate unless formal drainage is established.

There is debate comparing the established practice of open drainage and resection of dead tissue compared to repeated percutaneous drainage procedures or the percutaneous insertion of an indwelling catheter with irrigation. Authors disagree on the safety and efficacy of repeated aspiration. In view of this doubt it seems wise to tailor the approach to each case. It is likely that established infection and cavities should be treated by open drainage and intravenous antimicrobials. Small lesions, relatively recent infection and larger lesions in patients not well enough for open surgery may require repeated aspiration with or without indwelling catheter drainage. The value of instilling antimicrobials into the cavity is, really, unknown. The best evidence is that it is harmless but of limited efficacy. The critical steps are to establish the diagnosis, isolate the organism, drain the abscess and provide adequate systemic chemotherapy, usually by intravenous infusion.

Drainage

Catheters may be introduced by the same methods and approach as diagnostic aspiration [7–9] (Fig. 3.7). It is often helpful to use a guidewire inserted through a needle over which the catheter is placed. J-shaped guides reduce the risk of inadvertently withdrawing the wire when the needle is removed. There are a variety of types but they broadly consist of plastic tubes with multiple end holes to prevent blocking. Some are of single lumen and allow only closed drainage, they easily block and must be observed closely. Double-lumen catheters allow continuous irrigation with physiological fluids. They are less likely to block, but care should be taken in case the outward route blocks and saline is forced into adjacent cavities and tissue plane with no opportunity to escape. The catheter may be fixed by suture to the skin or by firm tape, taking care to avoid inflamed or necrotic tissue. Catheters may be of a pigtail shape to aid retention.

It is essential that the drainage is established and working properly before the patient is returned to the ward. Urine collection bags are ideal to provide a sealed

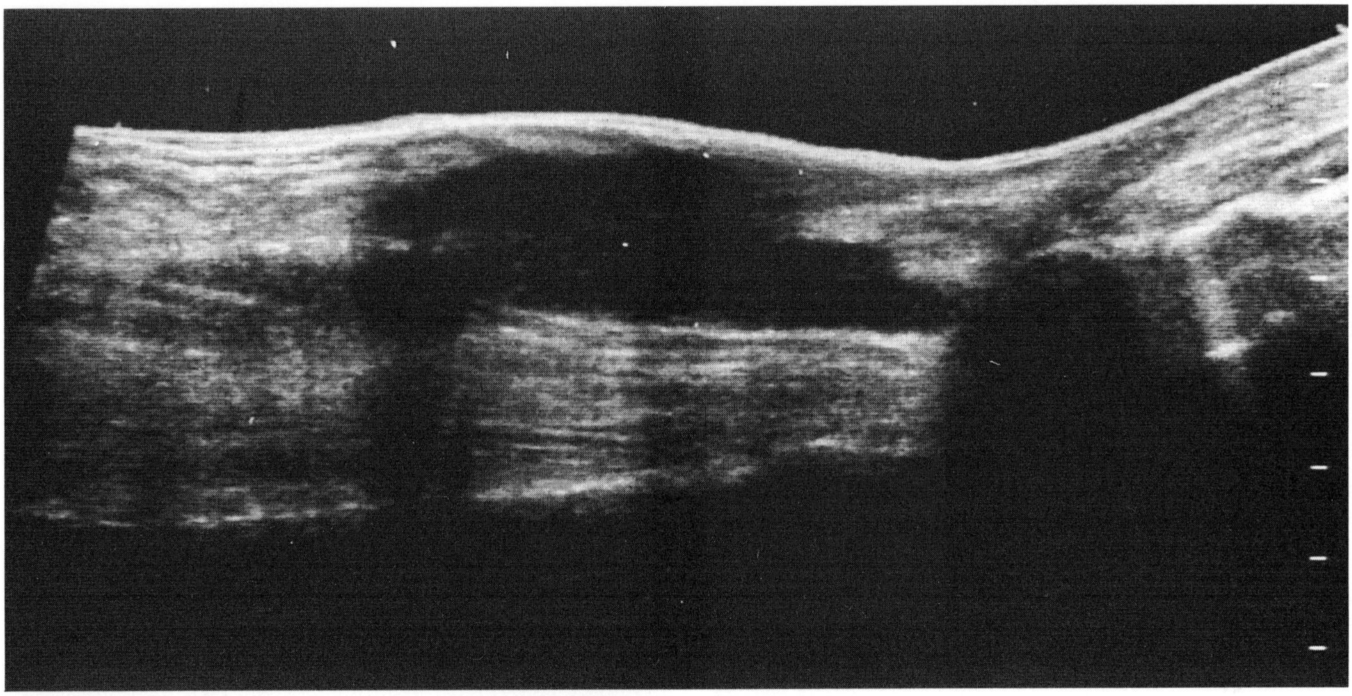

(a)

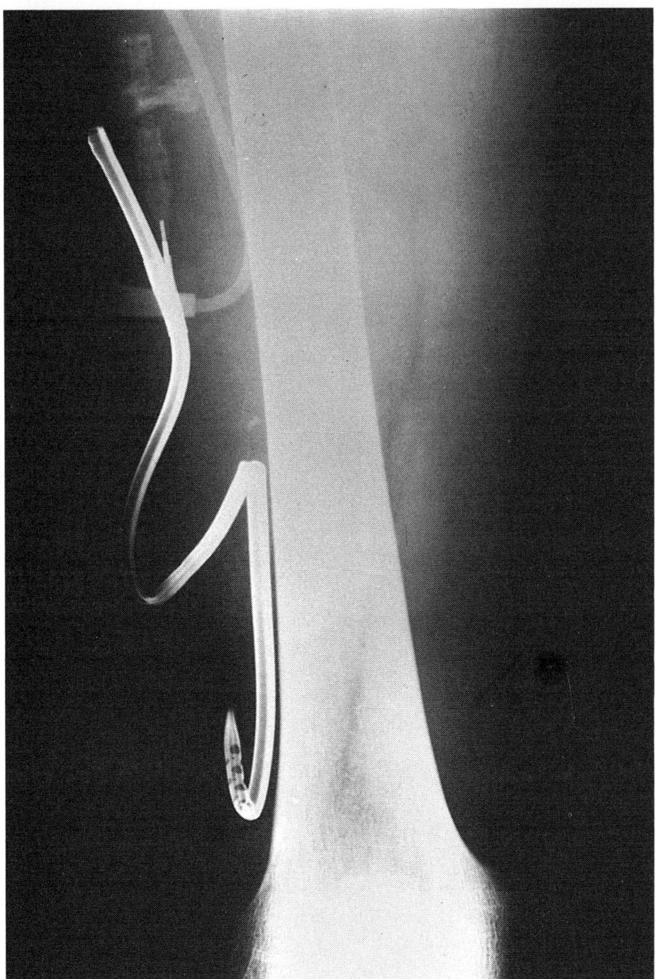

(b)

Fig. 3.7
(a) A composite ultrasound image of the thigh shows a large abscess with two main cavities. (b) A sump drainage catheter inserted under ultrasound control was threaded through both cavities. With intravenous antibiotics this therapy proved curative and open drainage was avoided.

route of drainage. The system should never be left open as additional infection is always a risk. The connecting tubes and adapters should be obtained before the procedure is undertaken. When drainage is by percutaneous methods it is vital to monitor the progress of the lesion by repeated ultrasound, CT, MRI or sinography. The frequency should be judged by progress and the size of the lesion although an early study in the day or two following placement is a useful baseline. The clinical team should be aware that open drainage may become necessary if the infection fails to respond rapidly. Irrigation with 5–20 ml of saline three to four times a day may aid drainage. The catheter may be removed when there is no more drainage but the imaging follow-up should be maintained. Less that 10 ml of drainage per day suggests that the abscess cavity is almost obliterated or that the catheter is blocked. Catheters should be changed after a few days if there is continued drainage and open drainage is rejected. Loculated collections may require additional percutaneous drainage or repeated aspiration. Small and superficial collections would normally drain within a week whilst larger and deep abscesses may take longer.[10] A number of reports have suggested that percutaneous drainage of psoas abscesses is very effective and probably the preferred method of treatment.[8, 11]

CONCLUSION

Imaging is the key to the diagnosis, staging, follow-up and management of deep sepsis.

REFERENCES

1 DALY BD, MCPHILLIPS M, LEUNG AW, EVANS RM, METREWELI C (1992). Ultrasound, computed tomography and magnetic resonance in the investigation of iliopsoas compartment disease. *Australasian Radiology* **36**(4): 294–9.

2 MALHOTRA R, SINGH KD, BHAN S, DAVE PK (1992). Primary pyogenic abscess of the psoas muscle. *Journal of Bone and Joint Surgery* **74**(2): 278–84.

3 MCGAHAN JP, BRANT WE, GERSCOVICH EO (1991). Ultrasound of the extremities and soft tissue masses. *Seminars in Orthopaedics* **6**(3): 145–55.

4 HUGHES DG, WILSON DJ (1986). Ultrasound appearances of peripheral nerve tumours. *British Journal Of Radiology* **59**: 1041–3.

5 MALHOTRA R, SINGH KD, BHAN S, .DAVE PK (1992). Primary pyogenic abscess of the psoas muscle. *Journal of Bone and Joint Surgery (American volume)* **74**(2): 278–84.

6 SCHMIDT RM, ROSENKRANZ HS (1970). Antimicrobial activity of local anaesthetics: Lidocaine and Procaine. *Journal of Infectious Diseases* **121**(6): 597–607.

7 GRUENWALD I, ABRAHAMSON J, COHEN O (1992). Psoas abscess: Case report and review of the literature. *Journal of Urology* **147**(6): 1624–6.

8 HELENON O, CORNUD F, DI STEFANO D, CHICHE JF, CHRÉTIEN Y, MOREAU JF, DUFOUR B (1989). Percutaneous treatment of abscess of the kidney and retroperitoneum/Traitement percutane des abcès du rein et du rétropéritoine. *Journal de Radiologie* **70**(10): 541–8.

9 CLEMENTSEN P, HANSEN M, CONRAD C, MYHRE O (1988). Percutaneous drainage of tuberculous abscess of the psoas muscle. *Tubercle* **69**(1): 63–5.

10 CHRISTENSEN RA, VAN SONNENBERG E, CASOLA G, WITTICH GR (1988). Interventional ultrasound in the musculoskeletal system. *Radiologic Clinics of North America* **26**(1): 145–56.

11 MUELLER PR, FERRUCCI JT, WITTENBERG J, SIMEONE JF, BUTCH RJ (1984). Iliopsoas abscess: Treatment by CT-guided percutaneous catheter drainage. *AJR* **142**: 359–362.

CHAPTER 4

Embolization of musculoskeletal lesions

James E. Jackson

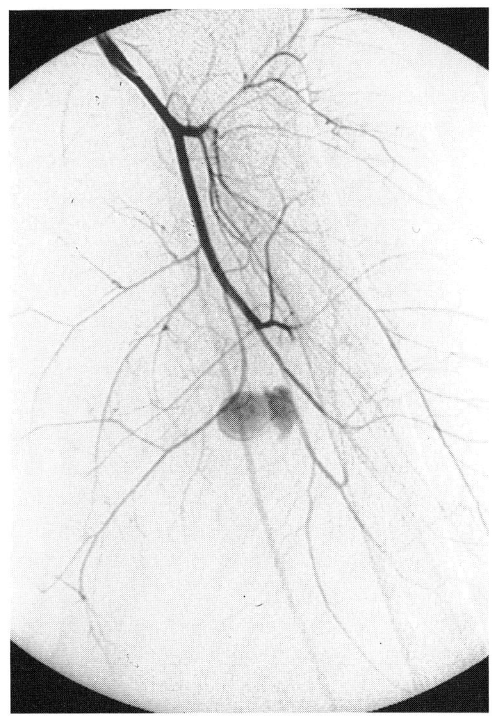

INDICATIONS

Vascular embolization may be extremely useful in the management of certain soft tissue and osseous lesions but is rarely performed outside specialist centres because of the relative rarity of these conditions. The indications for embolization are best considered under the following headings:

1 as a definitive treatment
2 pre-operatively to reduce vascularity or tumour size
3 for the palliation of symptoms due to inoperable tumours
4 for the management of haemorrhage following trauma.

Embolization as a definitive treatment

Vascular malformations involving soft tissue or bone are the musculoskeletal lesions most commonly treated by embolization. Embolization may be used alone or combined with surgery. It is obviously important that the true nature of these tumours is clearly understood, and a short description of their best classification (in the author's opinion) is given below.

Other neoplasms which may occasionally be successfully managed using embolization alone are aneurysmal bone cysts and, very rarely, infantile haemangiomas.

VASCULAR MALFORMATIONS

A review of the literature will show that the classification of vascular malformations is beset by a whole host of confusing descriptive terms, including angioma, pulsatile naevus, capillary or cavernous haemangioma and arteriovenous fistula. In the view of this author the use of a single method of classification is absolutely imperative, not least because it will lead to a better understanding of both the nature and the treatment of vascular malformations.

Perhaps the simplest and most workable classification is that of Mulliken.[1] A full description of this lies outside the scope of the present chapter and interested readers are strongly advised to refer to Mulliken's text. He suggests that the term 'haemangioma' should be restricted to a vascular tumour of infancy which passes through stages of rapid proliferation and subsequent spontaneous slow resolution. On histological examination such lesions have a plump endothelium and a multilaminated basement membrane. They show active capillary tubule formation *in vitro*.

A vascular birthmark, or vascular malformation, on the other hand, is always present at birth, although it may not become apparent until later life. Vascular malformations tend to grow commensurately with the patient, but an increase in growth rate may occur in response to certain stimuli such as trauma, puberty or pregnancy. They show no evidence of spontaneous involution. Histologically these lesions have a flat endothelium, a normal thin basement membrane and show poor endothelial growth *in vitro*.

Vascular malformations are subdivided according to the predominant type of vessel that is abnormal. This can usually be determined clinically. Lesions with a predominant arterial component are characterized by their pulsatile nature and rapid arteriovenous shunting. Lesions with a predominant venous component are non-pulsatile, have normal feeding arteries and consist of large ectatic channels which often contain phleboliths.

The method of treatment depends upon their location, associated symptoms and on the predominant vessel type. Very few malformations can be definitively treated by embolization alone, although in rare cases a cure can be achieved even when the lesion is large. Slightly more malformations may be cured using a combination of embolization and surgery but, unfortunately, by far the majority of these lesions can only be palliated by embolization because of their large size and diffuse nature.

Curative embolization may be attempted in lesions that are relatively small and localized so long as embolization of adjacent normal territory can be avoided (Fig. 4.1). Complete ablation of the lesion is required in such instances and this is often best achieved with a powerful sclerosant such as absolute alcohol. Such agents must be used with extreme care because of their potential to cause catastrophic complications (see below under 'Embolic agents').

An exception to the rule that large malformations cannot be cured by embolization alone appears to be those, seen in men, which are fed by the internal iliac artery. In one case treated by the author[2] and two others treated at another institution (R. J. Rosen, New York University Medical Centre, personal communication) a cure appears to have been achieved by embolization using 'conventional' embolic agents.

INFANTILE HAEMANGIOMAS

Embolization (or indeed any active therapy) of haemangiomas is rarely required. Active treatment is unnecessary in most cases, although there are exceptions. Haemangiomas which are particularly large, those that cause severe haemorrhage because of platelet trapping and subsequent profound thrombocytopaenia (Kasabach–Merritt syndrome), and those that cause complications owing to their position (e.g. obstruction of the visual axis) may require treatment.

ANEURYSMAL BONE CYSTS

Aneurysmal bone cysts are generally considered benign. Trauma may play an aetiological role in some cases and they may occasionally accompany bone neoplasms such

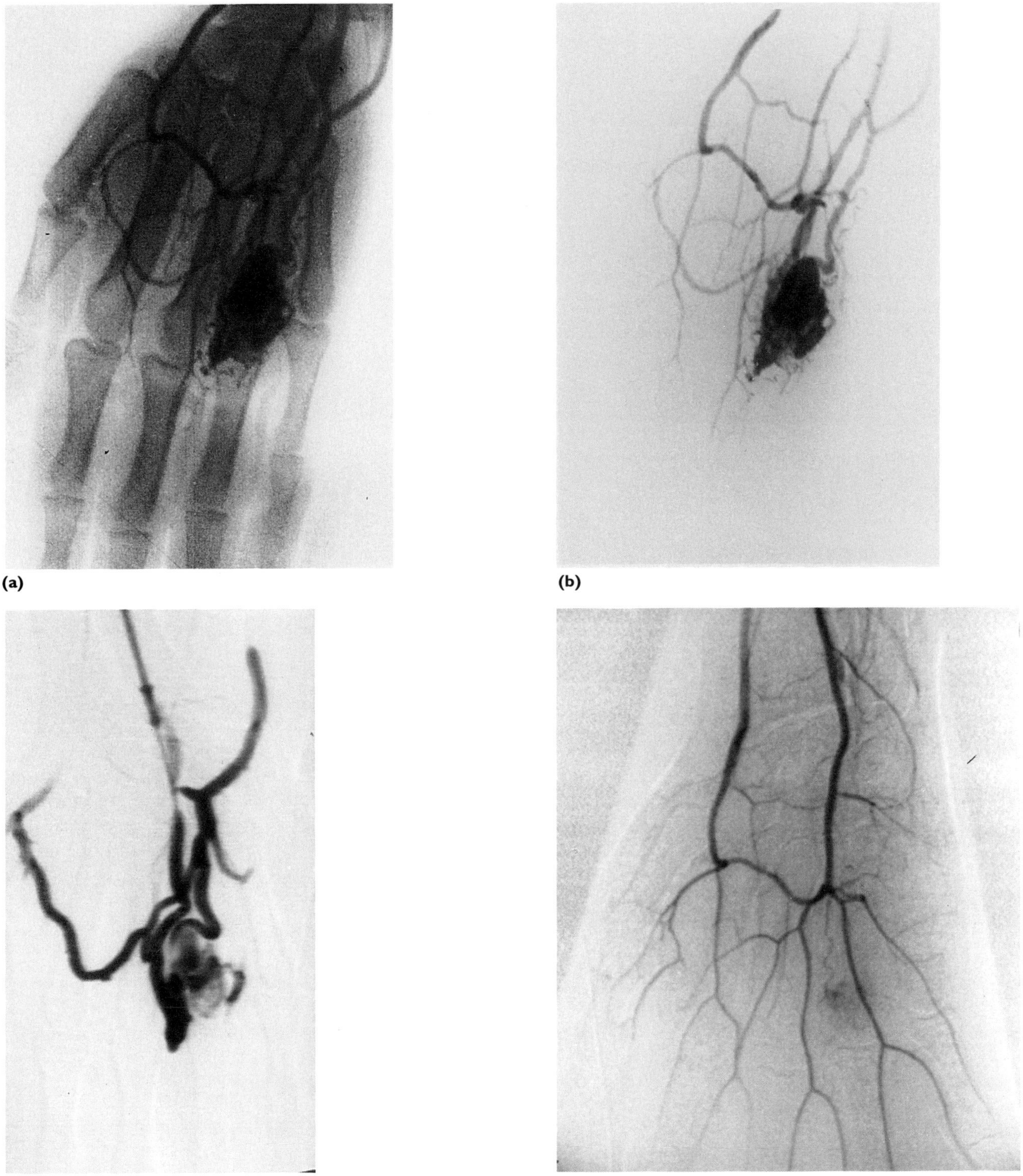

(a) (b) (c) (d)

Fig. 4.1
Localized hand vascular malformation treated via a retrograde venous approach with absolute alcohol. (a, b) Left hand arteriogram performed using a digital subtraction technique. A localized high-flow malformation is present over the dorsum of the hand with rapid arteriovenous shunting. Note the tortuous arterial supply and the proximity to normal digital vessels, making arterial embolization difficult and hazardous. (c) One of the major draining veins has been punctured over the dorsum of the wrist. A venogram performed during inflation of a tourniquet demonstrates good retrograde filling of the malformation. A co-axial catheter was passed retrogradely into the centre of the malformation and embolization was performed with absolute alcohol. (d) An arteriogram performed 3 months later demonstrates almost complete obliteration of the malformation. The patient remains well 2 years following the embolization with no evidence of recurrence.

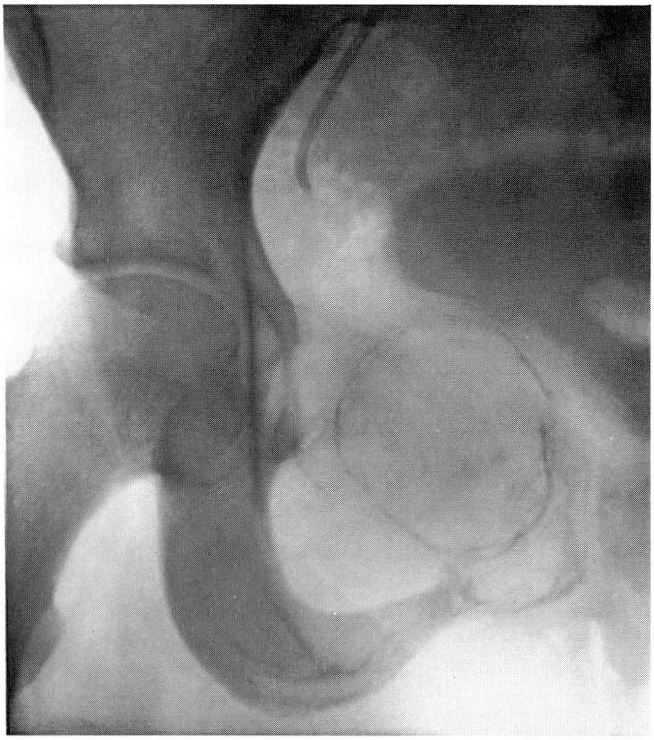

(a)

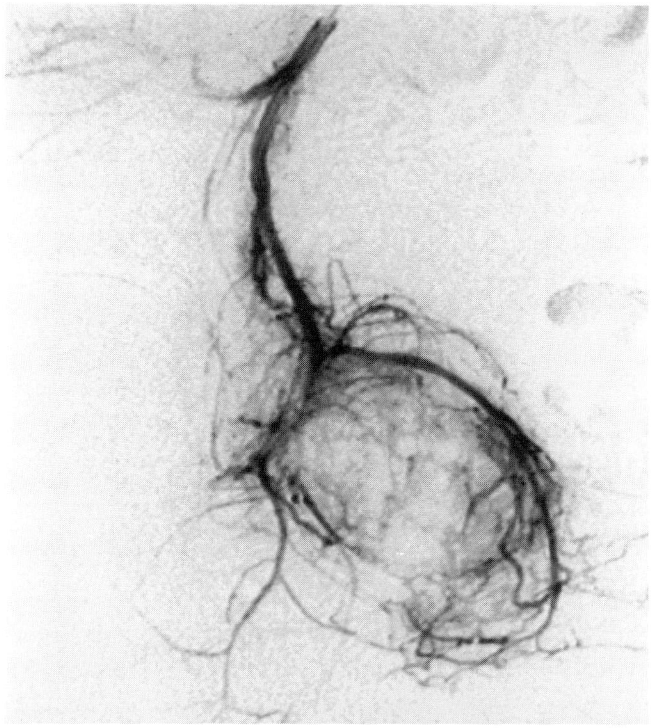

(b)

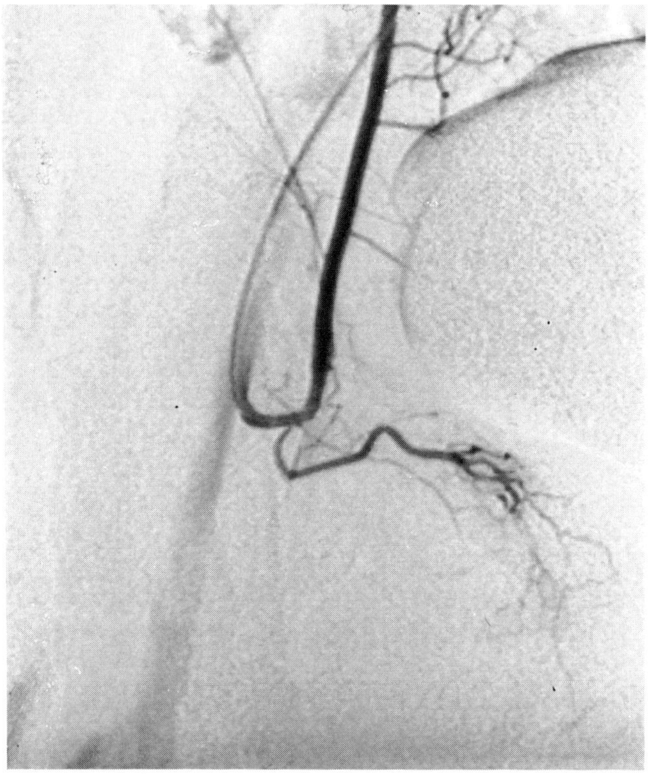

(c)

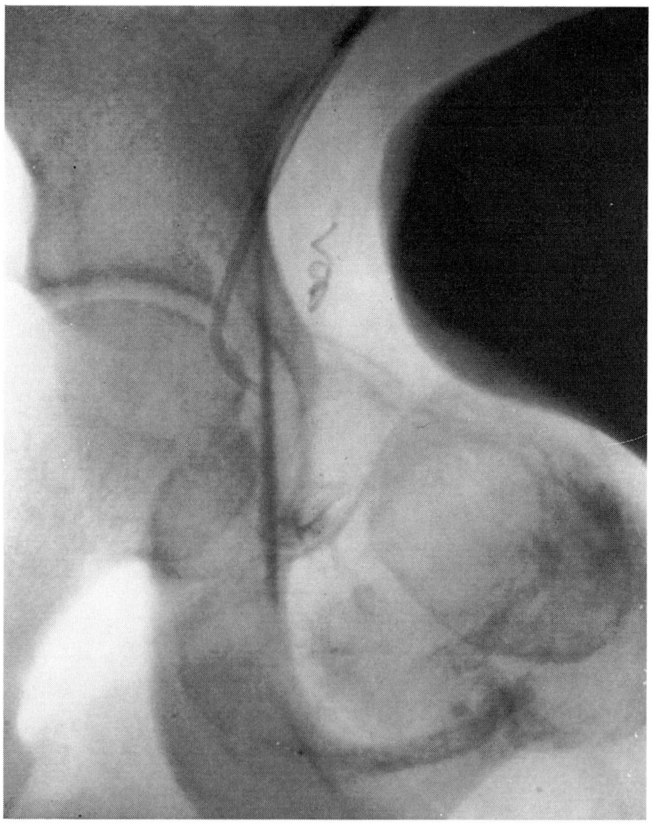

(d)

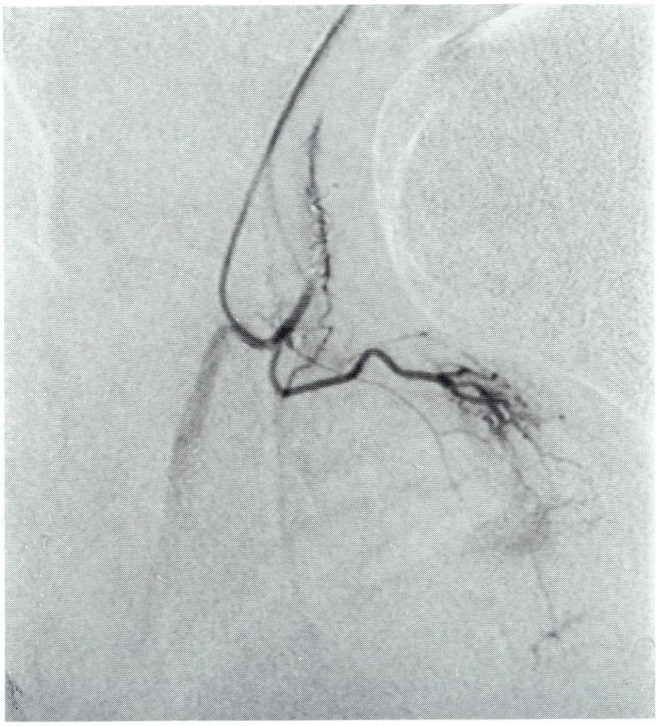

(e)

as giant cell tumours, osteoblastomas and chondromyxoid fibromas.[3] Although there is no risk of metastasis, these tumours may grow rapidly with considerable destruction of bone and extension into adjacent soft tissue, and aggressive treatment is often undertaken. Excellent results may, however, be obtained by embolization alone (Fig. 4.2), and when surgery is very difficult or impossible this form of therapy may control and sometimes cure the disease.[4]

Preoperative embolization

Surgical excision of musculoskeletal tumours is often difficult because of their vascularity. Preoperative embolization of both benign and malignant vascular soft tissue and bone lesions may be very helpful. This is especially so for surgical resection of soft tissue vascular malformations. It is important to realize that the primary objective of preoperative embolization is to reduce the vascularity of the lesion and not to reduce the extent of surgical excision. To acheive a permanent cure of a vascular malformation, complete excision of the lesion is essential, but is not always possible. Any other vascular

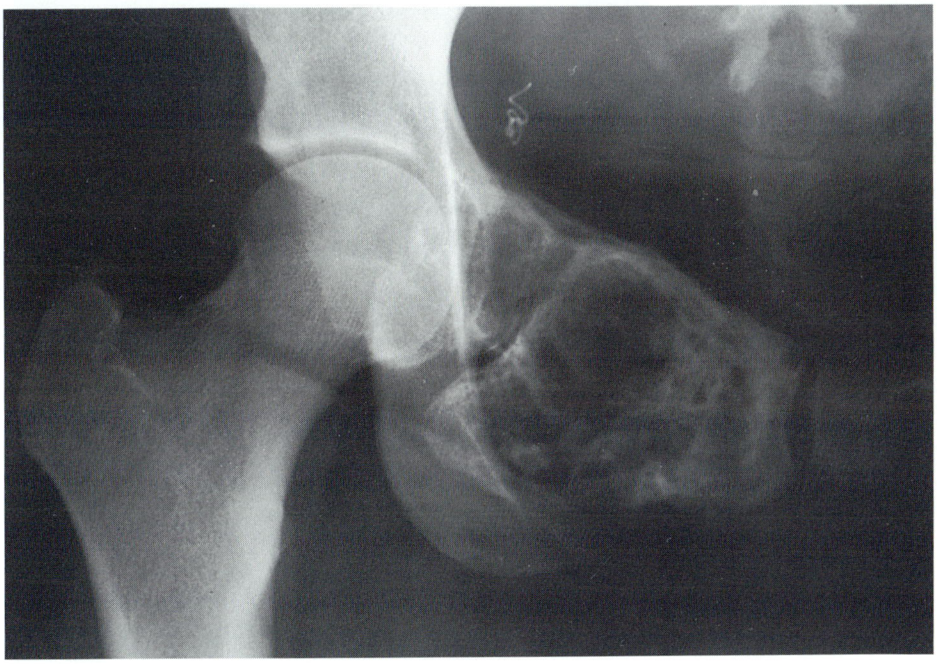

(f)

Fig. 4.2
Aneurysmal bone cyst of the right inferior pubic ramus treated by embolization. (a) A plain radiograph demonstrates a very expanded and destructive lesion involving the right inferior pubic ramus. Biopsy confirmed the radiographic diagnosis of an aneurysmal bone cyst. (b) Selective right obturator artery angiogram using a digital subtraction technique demonstrates the vascularity of the lesion, which also obtained some supply from the inferior epigastric artery and from the circumflex femoral vessels. (c) A right inferior epigastric arteriogram performed during a second embolization demonstrates some residual supply to the tumour from this vessel. (d) A metal coil has been introduced into the inferior epigastric artery beyond the vessels supplying the tumour to protect the normal vascular bed. (e) An angiogram demonstrates occlusion of the inferior epigastric artery. Embolization of the tumour vessels was then performed with polyvinyl alcohol. (f) Plain film 1 year following the second embolization shows extensive reossification and bone remodelling. The patient remains asymptomatic 2 years after initial presentation.

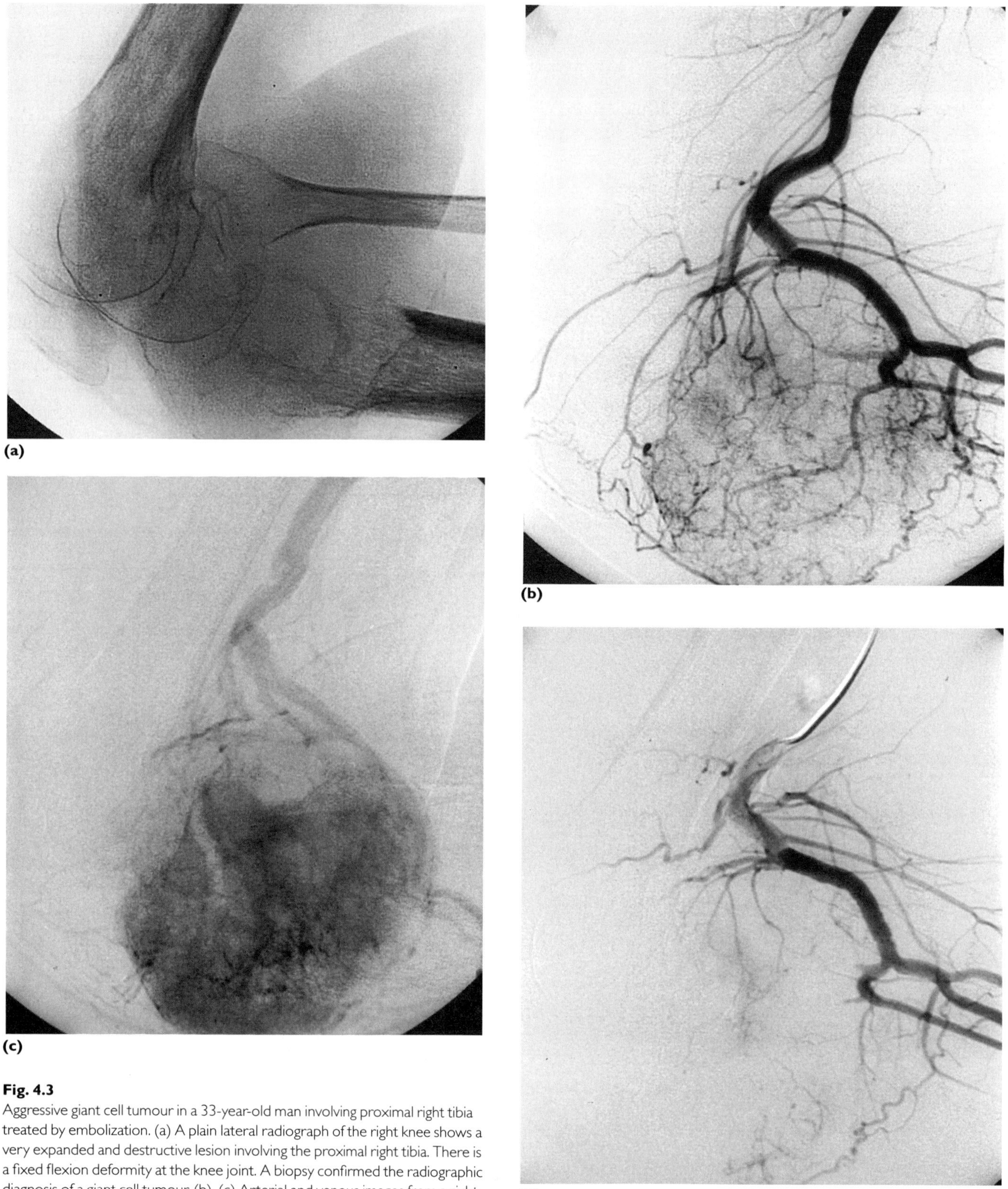

(a) (b) (c) (d)

Fig. 4.3
Aggressive giant cell tumour in a 33-year-old man involving proximal right tibia treated by embolization. (a) A plain lateral radiograph of the right knee shows a very expanded and destructive lesion involving the proximal right tibia. There is a fixed flexion deformity at the knee joint. A biopsy confirmed the radiographic diagnosis of a giant cell tumour. (b), (c) Arterial and venous images from a right femoral arteriogram demonstrate that the lesion is extremely vascular and derives its supply from numerous branches arising from the popliteal, anterior tibial and tibio-peroneal arteries. (d) Right femoral arteriography post-embolization shows a marked reduction in tumour vascularity. Some supply remains from the tibio-peroneal trunk. This procedure produced a significant reduction in soft tissue swelling and marked symptomatic improvement. Further embolization is planned before definitive surgery.

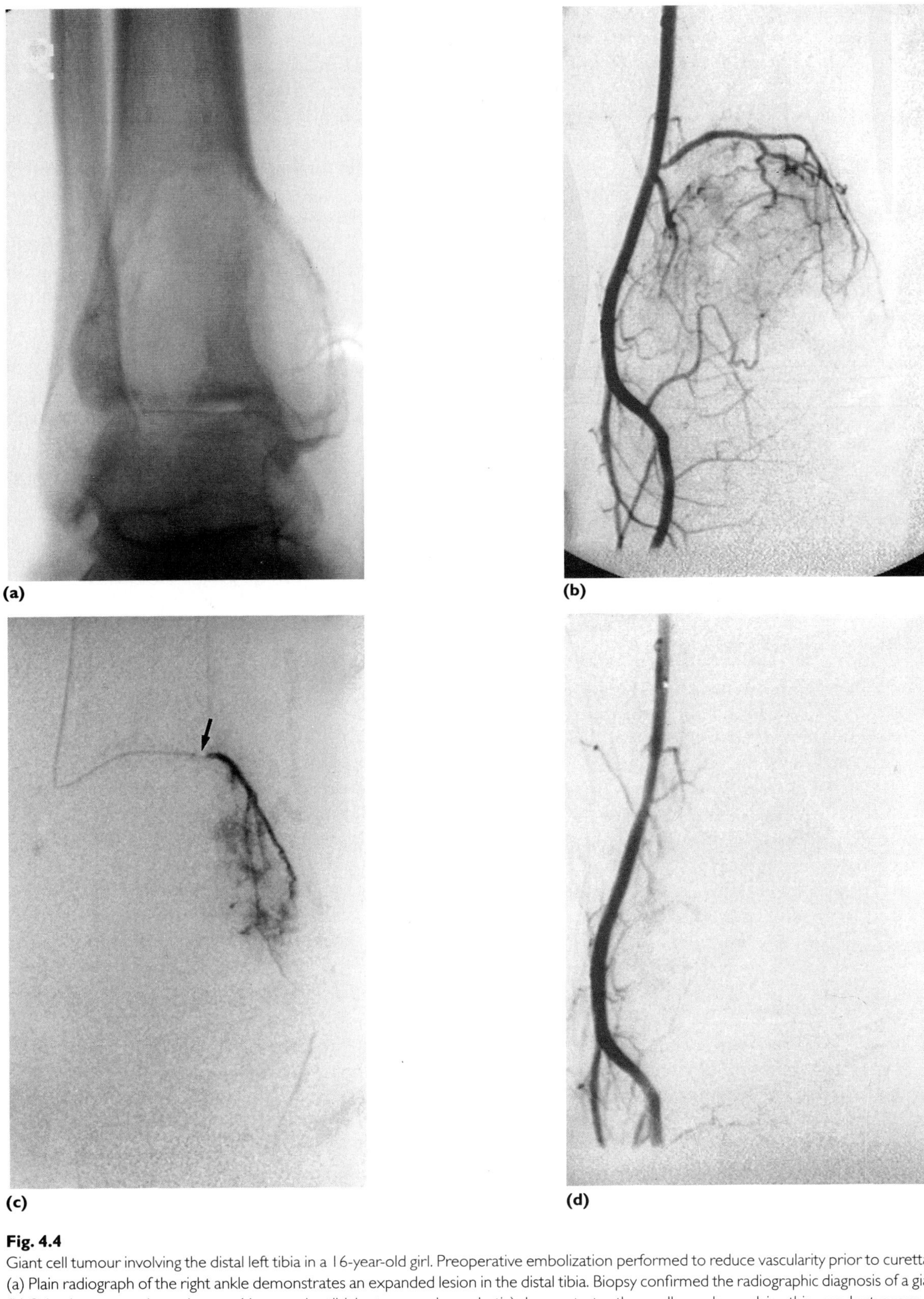

Fig. 4.4
Giant cell tumour involving the distal left tibia in a 16-year-old girl. Preoperative embolization performed to reduce vascularity prior to curettage and cementation. (a) Plain radiograph of the right ankle demonstrates an expanded lesion in the distal tibia. Biopsy confirmed the radiographic diagnosis of a giant cell tumour. (b) Selective peroneal arteriogram (the anterior tibial artery was hypoplastic) demonstrates the small vessels supplying this vascular tumour. (c) Superselective catheterization of one of the feeding vessels arising from the distal peroneal with a co-axial 3 Fr catheter (arrow) prior to embolization of this branch. (d) Post-embolization peroneal arteriogram demonstrates a marked reduction in tumour vascularity (compare with Fig. 4(b)). The patient subsequently underwent curettage and cementation (filling of the tumour cavity with bone cement).

soft tissue tumour (e.g. haemangiopericytoma, alveolar soft part sarcoma, soft tissue metastasis) may benefit from preoperative embolization in an attempt to reduce blood loss at surgery.

Perhaps the most common (yet still rare) primary *osseous* neoplasm which may benefit from preoperative embolization is the giant cell tumour when it is particularly aggressive or advanced at presentation. These more rapidly growing tumours are more common in men (M : F = 3 : 1) and may be difficult to treat surgically because of their large size and the extensive bone destruction (Fig. 4.3). Embolization, perhaps repeated on several occasions a few weeks or months apart, can produce both a marked reduction in tumour size and considerable regrowth and remodelling of bone, thus allowing more conservative surgery.[5] Because of the high recurrence rate of giant cell tumours following surgery, and the occasional report of malignant transformation, operative treatment is still recommended (Fig. 4.4) even if the response to embolization is dramatic.

Other vascular bone tumours, in particular metastases from renal cell carcinoma, may be very usefully treated by embolization in order to prevent life-threatening bleeding during subsequent surgical fixation.

TIMING OF SURGERY

For the majority of soft tissue and osseous lesions surgical excision or bony fixation should, ideally, be performed within 48 hours of embolization. Thrombosis of the lesion will have occurred by this time and the surrounding fat planes will still be preserved. If excision is attempted at a later date (1–2 weeks), inflammation of the surrounding soft tissue, fibrosis and loss of the normal fat planes will make surgery more difficult. In addition, a significant arterial supply may have returned to some of the more vascular and aggressive lesions even at this relatively early stage. Larger lesions may benefit from two or more embolization procedures spaced several weeks apart, with surgery being performed within 48 hours of the final embolization.

In the case of lesions such as aggressive giant cell tumours, where repeated embolization is being performed to produce a reduction in tumour size and bone remodelling, the timing of surgery will depend upon the tumour's response to treatment and may be delayed for several weeks or months (Fig. 4.3).

Palliative embolization

The most common indication for palliative embolization in the musculoskeletal system is in the treatment of vascular malformations whose size or location means that curative resection would be either too hazardous or too disfiguring. In such cases embolization is aimed at the alleviation of symptoms and often achieves this satisfactorily, at least in the short term (Fig. 4.5). It should be remembered that these lesions are benign, that symptoms are often not particularly severe and that, in the light of present knowledge and available embolization agents, most are incurable. It is, therefore, extremely important that over-zealous embolization is not attempted as there may be disastrous complications. There can be few things worse than attempting to treat a relatively asymptomatic peripheral malformation by embolization and leaving the patient with a useless limb or precipitating amputation.

Pain and bleeding from inoperable primary or secondary musculoskeletal tumours may be well controlled by vascular embolization. Expanded bony metastases, from renal cell carcinoma for example, may respond extremely well to embolization, and good palliation may be achieved even in those patients in whom other forms of more conventional therapy have failed.[6]

Management of haemorrhage following trauma

In major trauma centres the control of haemorrhage, particularly in the pelvis, is one of the major indications for embolization and is achieved by selective occlusion of the bleeding vessel if this is visualized or of one or both internal iliac arteries (for pelvic haemorrhage) if it is not. An example of its use in a patient with a stab wound to the thigh is shown in Fig. 4.6.

METHODS

The approach used for embolization depends upon the nature of the lesion being treated. Occlusion of the arterial supply is the most commonly performed procedure but a direct percutaneous approach into the tumour (particularly those lesions which are predominantly venous) or a retrograde venous route may be more appropriate in certain instances. The techniques described below relate predominantly to arterial embolization but the basic principles apply to embolization of any sort.

1. A thorough knowledge of the normal local vascular anatomy is essential.
2. It is imperative that a complete angiographic study is performed before embolization in order to define the full extent of the lesion being treated and to demonstrate all of the potential feeding vessels regardless of the vascular territory it involves. For example, when

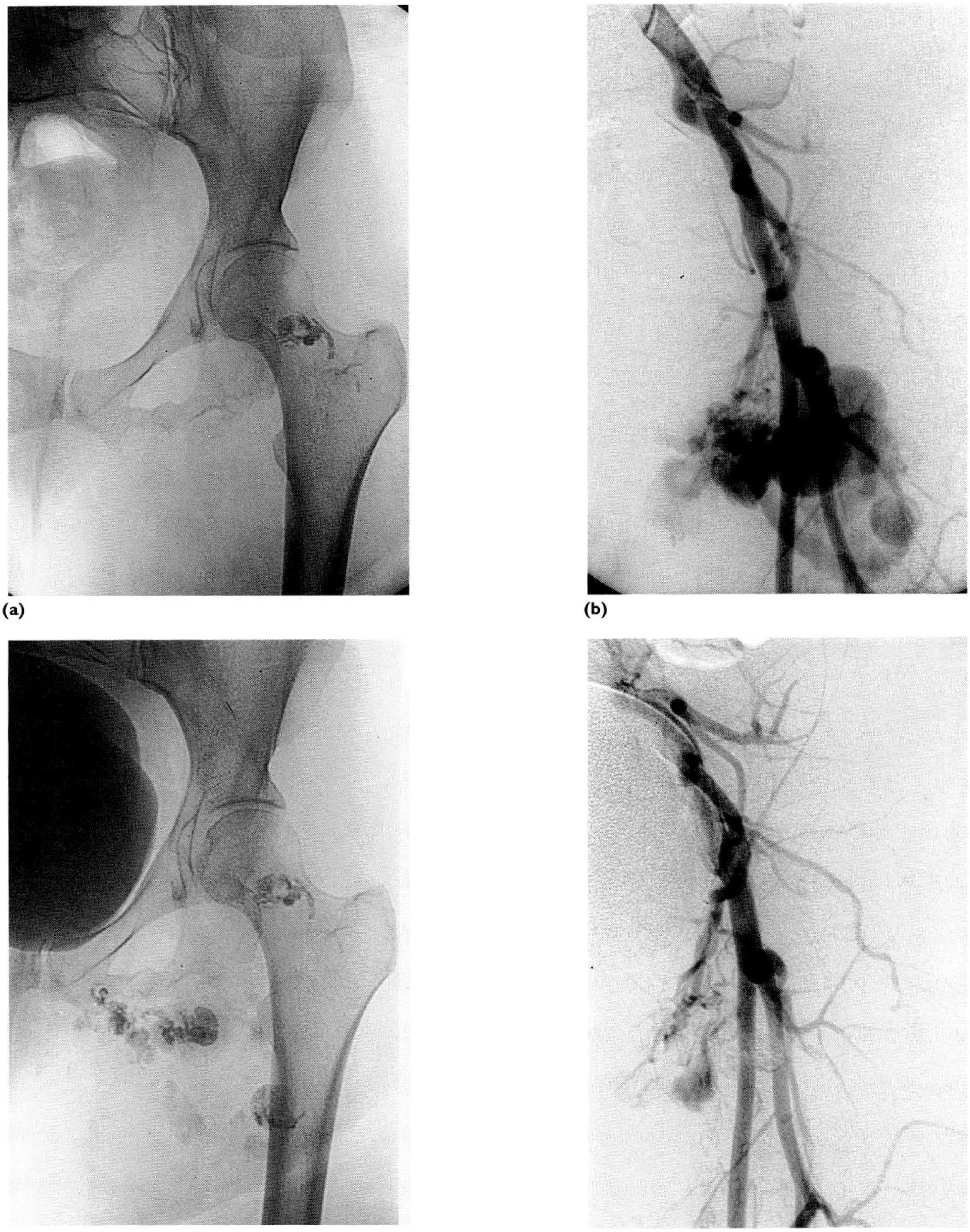

(a) (b) (c) (d)

Fig. 4.5
Massive high-flow vascular malformation involving the left thigh treated by embolization with isobutyl-2-cyanoacrylate (IBCA). (a) Plain radiograph of left pelvis and upper thigh. Note the bony changes within the left inferior pubic ramus due to pressure erosion by the vascular malformation. The density overlying the left femoral head is IBCA from a previous embolization. (b) Early arterial image from a left common iliac arteriogram demonstrates a massive vascular malformation with rapid arteriovenous shunting into aneurysmally dilated draining veins. (c) Plain radiograph following embolization shows the IBCA (mixed with iodized oil) overlying the left upper thigh. (d) Left common iliac arteriogram post-embolization shows the marked reduction in arteriovenous shunting that has been achieved.

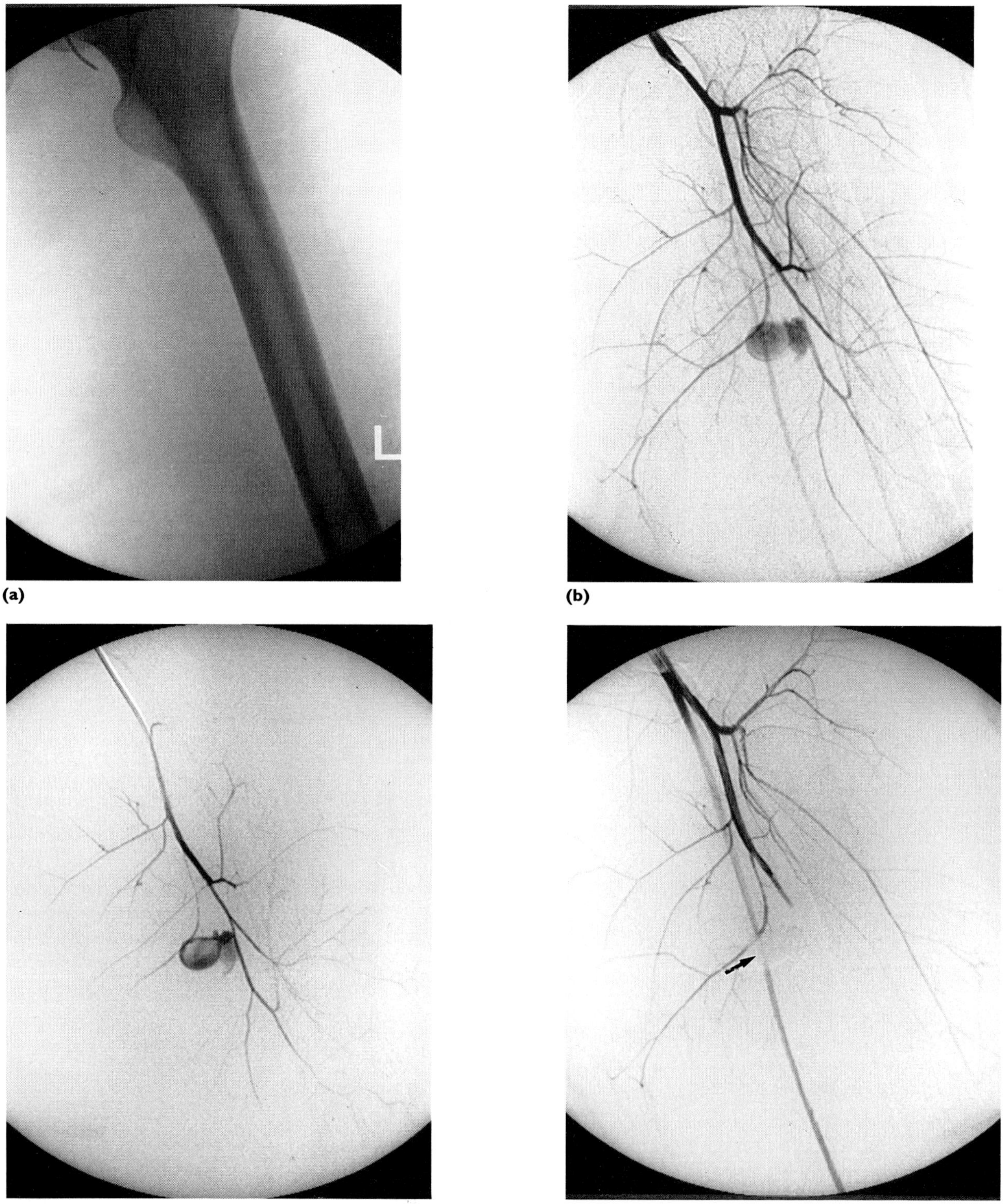

(a) (b) (c) (d)

Fig. 4.6
Stab wound to left thigh with severe haemorrhage uncontrolled by surgery. (a), (b) Left femoral arteriogram demonstrates brisk contrast medium extravasation from a branch of the profunda femoris. (c) Selective catheterization of the feeding vessel demonstrates the arterial defect. (d) Following embolization with absorbable gelatin sponge the contrast medium extravasation is no longer seen. The patient had no further bleeding. The apparent defect in the superficial femoral artery (arrow) is an artefactual appearance caused by the overlying extravasated contrast medium.

embolizing lesions supplied predominantly by the external carotid artery, bilateral external carotid and ipsilateral internal carotid artery angiograms are mandatory. An ipsilateral vertebral artery angiogram is also necessary if the lesion extends posterior to the external auditory meatus. An internal carotid artery supply to a vascular facial lesion (often via the ophthalmic artery) is not uncommon and, unless the presence of such a supply is appreciated, potentially catastrophic complications may occur owing to the retrograde passage of embolic material into these vessels.

3 The changes in blood flow which occur in the territory being embolized and in the surrounding normal vascular bed during vessel occlusion must be appreciated.

4 Central occlusion of the lesion should be aimed for by using the smallest particle size that does not pass through from the arterial into the venous circulation. This produces the best clinical result by preventing the rapid re-perfusion of the lesion by collateral vessels.

5 Proximal coil or balloon embolization of the feeding vessels to the lesion must be avoided as this manoeuvre precludes successful embolization of the lesion at a later date should this be required. This is especially true for vascular malformations as repeated embolization is frequently necessary, but should also apply to lesions which are undergoing surgical excision as the presence of coils or balloons within proximal vessels will make embolization considerably more difficult and less successful should the tumour recur.

6 A selective catheter position must be achieved beyond vessels supplying any normal vascular territory adjoining the malformation. Co-axial catheters (small catheters which pass through the central lumen of a conventional angiographic catheter) are invaluable in certain vascular territories (Figs 4.4d and 4.4e).

7 Embolic agents should always be mixed with a contrast agent so that they are visible during injection.

8 The embolic agent being used must be injected in small amounts and under continuous fluoroscopy so that the inadvertent embolization of a normal vascular territory is avoided. The use of a 1- or 2-ml syringe is advised (particularly towards the end of the embolization) as this makes it considerably easier to control the rate of injection, thus preventing the overspill of embolic agents into normal proximal vessels.

9 Repeated contrast medium injections and angiographic images should be obtained in order to monitor continuously the progress of the embolization between successive injections of small amounts of embolic material. The use of digital subtraction angiographic (DSA) equipment makes embolization considerably safer as repeated angiograms may easily be performed during vessel occlusion. It may not be obvious during fluoroscopy that, as the abnormal vascular bed is being occluded, some spill of embolic material is occurring into normal vessels because rapid blood flow very quickly dilutes the contrast medium. A short DSA run with its superior contrast resolution allows one to monitor the progress of embolization easily and rapidly, and will clearly demonstrate whether or not any overspill is occurring.

10 Tumours often obtain a parasitic blood supply from surrounding organs and this may make complete embolization very difficult, if not impossible. If it is planned to remove the normal organ which is contributing its blood supply to the tumour at subsequent surgery, then embolization of all or part of this organ may also be performed in order further to decrease the vascularity of the neoplasm. If there is a possibility that an attempt will be made to preserve the normal organ at surgery, then embolization is inappropriate even if this means that some arterial supply to the tumour remains.

Catheters

The variety of catheters now available means that there are few vessels which cannot be successfully catheterized selectively. Co-axial catheters which pass down the centre of a conventional angiographic catheter are particularly useful in certain circumstances and may allow safe embolization of lesions that would not previously have been amenable to this form of treatment. The refinement of these co-axial techniques in recent years has revolutionized the treatment of certain lesions, particularly those involving intracranial or extremity (hand and foot) vessels where preservation of surrounding normal arteries is especially important (Fig. 4.4(d)) and 4.4(e)). These catheters generally have a 3 Fr proximal shaft and taper distally to about 2.2 Fr. Their small internal diameter (0.018 in., 0.46 mm) allows the passage of small particulate embolic agents up to about 500 μm in diameter (see below under 'Embolic agents').

Embolization of many soft tissue or bony lesions, however, may be performed using conventional 5 or 7 Fr catheters. The configuration of catheter chosen will obviously depend upon the vascular territory being embolized; in the author's experience the Headhunter, Cobra and Sidewinder shaped catheters are those most usefully employed. It is important to use an end-hole only catheter when performing embolization. Proximal side-holes increase the likelihood of inadvertent embolization of an adjacent or distant normal vascular territory. An 0.038-in. (0.97-mm) diameter lumen is recommended as this not only eases the injection of the embolic material, but also allows the passage of a co-axial catheter should this become necessary.

Embolic agents

A variety of embolic agents are available. Those most commonly used are as follows:

Particulate agents
- Polyvinyl alcohol (PVA)
- Absorbable gelatin sponge
- Metal coils
- Detachable balloons

Liquids
- Dehydrated alcohol
- Sodium tetradechol
- Isobutyl-2-cyanoacrylate (IBCA)
- Hypertonic dextrose

The type of embolic agent used in any particular situation depends upon the lesion being embolized and the indication for the procedure. Whatever the type of lesion being embolized, it is important to try to occlude the central tumour vessels as this produces the best result in terms of tissue ischaemia. If only proximal vessel occlusion is performed, then, although an arteriogram at the end of the procedure may show no perfusion of the lesion, collateral vessels will soon develop (often within 24 hours). In such cases, if the embolization was being performed prior to surgery then there are likely to be more problems with haemostasis than if central vessel occlusion had been performed. If the procedure was being performed for palliation then the rapid development of collaterals will mean that there is unlikely to be satisfactory control of symptons in the short or long term.

Arguably the best available agent in the majority of circumstances for embolization of vascular musculoskeletal tumours is polyvinyl alcohol (PVA). This agent is available in dry powder form in a variety of graded particle sizes, the selection of which may therefore be tailored to the size of the central vessels (or arteriovenous communications) in the lesion being treated. PVA is said to be a permanent embolic agent but, in the author's experience and that of others, recanalization of previously occluded vessels is not uncommon.[7]

Absorbable gelatin sponge is a temporary embolic agent and produces arterial occlusion for between several days and several months. It is available as small sheets which are then cut to the desired size prior to injection. It may be mixed with a contrast agent to form a 'slurry', or 'single torpedoes' of the sponge may be injected using a 1- or 2-ml syringe. It is generally used for occlusion of larger vessels and is particularly useful in trauma where it is used to occlude haemorrhaging vessels. As mentioned previously, coils and detachable balloons should generally be avoided when embolizing musculoskeletal tumours as they prevent access to the lesion at a later date should embolization be required again. They may occasionally be used, however, to protect a normal vascular bed from embolization with small particulate matter (Figs 4.2(f) and 4.2(g)). Metal coils may interfere with subsequent MR imaging employed to assess recurrent disease.

Liquid embolic agents such as absolute alcohol, sodium tetradechol (STD) and isobutyl-2-cynanoacrylate (IBCA) are extremely irritant and must be used with extreme caution. There is no doubt that these agents may produce quite dramatic results in some cases[8] (see Fig. 4.1), but serious complications such as skin and muscle necrosis, neural damage and infarction of normal adjacent or distant organs are not uncommon. There is certainly little or no place for their use in the majority of lesions being embolized pre-operatively because the particulate agents, which are in any case much easier to handle, will produce as good a result. IBCA may produce good results in high-flow arteriovenous malformations which are not amenable to surgical excision (Fig. 4.5).

The one situation where the liquid agents STD and absolute alcohol have been shown to be particularly useful is in the treatment of venous malformations. These lesions are directly punctured and venography is performed to document their size and to confirm that there is no spillover into normal veins. Sclerotherapy is then performed with STD or absolute alcohol. Several punctures may be required into different portions of the malformation as it is unusual to be able to fill the entire lesion from a single site. Satisfactory results can be obtained but complete obliteration of the lesion is unlikely.

Complications

Complications directly attributable to embolization are fortunately extremely uncommon. Those most commonly reported during embolization of musculoskeletal lesions are the inadvertent embolization of vessels other than those supplying the tumour and, in the case of vascular malformations, the passage of emboli into the venous circulation.[9, 10] Provided the angiographer has a thorough knowledge of the normal vascular anatomy in the territory being embolized and the embolization procedure is performed with meticulous care, these complications should be very uncommon.

Embolic agents such as coils and detachable balloons may very occasionally be misplaced and thereby occlude an important normal vessel. It is usually possible for the angiographer to remove or reposition these devices percutaneously with an appropriate retrieval device, but if this is not achieved then surgery may be necessary.

The post-embolization syndrome (PES), which consists of a variable combination of pain at the site of the lesion which has been embolized, pyrexia, leucocytosis and a general feeling of malaise, is common and may last between 24 hours and 2 weeks. The more serious complication of infection at the site of embolization is exceedingly rare but should always be considered, particularly in the case of a severe or unduly prolonged PES. Blood cultures should be obtained in any case featuring pyrexia and leucocytosis.

Tissue necrosis is fortunately also an infrequent complication of embolization but is more likely if liquid sclerosants such as alcohol or IBCA are used.

CLINICAL VALUE

There can be little doubt that in certain situations embolization has a major part to play. It may be life-saving in the case of severe haemorrhage following trauma, particularly when this involves the pelvis, where operative localization and control of the bleeding vessel may be impossible.

In the case of a vascular soft tissue or osseous tumour it may make surgical excision or internal fixation of bone easier by reducing the vascularity of the lesion. In certain individuals, for example patients with inoperable aneurysmal bone cysts or soft tissue or bone vascular malformations, embolization may be the only form of therapy available. The latter lesions are extremely difficult to treat and cure is unlikely, but significant symptomatic improvement can be obtained in most cases.

REFERENCES

1 MULLIKEN JB (1988). Classification of vascular birthmarks. In: *Vascular Birthmarks, Haemangiomas and Malformations*, pp. 24–37. Edited by Mulliken JB, Young AE. W.B. Saunders Company, Philadelphia.

2 JÄGER HR, JACKSON JE, ALLISON DJ (1993). Delayed pulmonary embolism after therapeutic vascular embolization of an arteriovenous malformation: Treatment with a venous filter. *Journal of Interventional Radiology* **7**: 153–6.

3 RESNICK D, KYRIAKOS M, GREENWAY GD (1988). Tumors and tumor-like lesions of bone: Imaging and pathology of specific lesions. In: *Diagnosis of Bone and Joint Disorders*, 2nd edition, pp. 3831–42. Edited by Resnick D, Niwayama G. W.B. Saunders Company, Philadelphia.

4 SUBY-LONG T, BOS GD, RÖSCH J (1988). Biopsy proven eradication of an aneurysmal bone cyst treated by superselective embolization: A case report. *Cardiovascular and Interventional Radiology* **11**: 292–5.

5 ALLISON DJ, HEMINGWAY AP (1987). Tumour embolization. In: *Bone Tumour Management*, pp. 75–81. Edited by Coombes R, Friedlaender G. Butterworth & Co., London.

6 CHUANG VP, WALLACE S, SWANSON D *et al.* (1979). Arterial occlusion in the management of pain from metastatic renal carcinoma. *Radiology* **133**: 611–14.

7 MILLER DL (1987). Failure of Ivalon to provide permanent hepatic arterial occlusion. *Cardiovascular and Interventional Radiology* **10**: 111–13.

8 YAKES WF, HAAS DK, PARKER SH *et al.* (1989). Symptomatic vascular malformations: Ethanol embolotherapy. *Radiology* **170**, 1059–66.

9 NOVAK D (1990). Complications of arterial embolization. In: *Interventional Radiology*, pp. 314–24. Edited by Dondelinger RF, Rossi P, Kurdziel JC, Wallace S. Thieme Medical Publishers, New York.

10 ALLISON DJ, WALLACE S, MACHAN LS (1992). Interventional radiology. In: *Diagnostic Radiology. An Anglo-American Textbook of Imaging*, 2nd edition, pp. 2329–90. Edited by Grainger RG, Allison DJ. Churchill Livingstone, Edinburgh.

3

Biopsy

CHAPTER 5

Bone biopsy

Philip Gishen and Fiona M. Witham

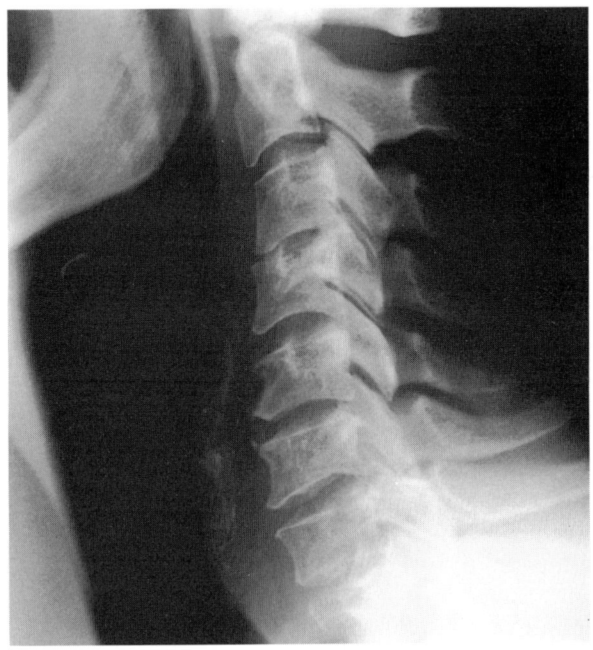

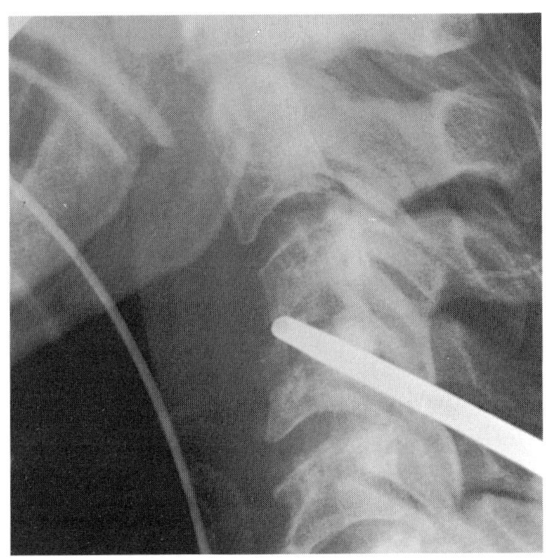

Introduction

In many patients presenting with bone problems, the clinical history, examination and radiological studies should indicate the correct diagnosis, but it is often preferable to have a histological diagnosis or to isolate an organism before commencing antibiotic therapy. Diagnosis of bone lesions by percutaneous needle biopsy has been carried out at least since 1926 when Martin and Ellis,[1] using 18 G 5–10 cm needles, aspirated the soft tissue components of bone tumours. Coley *et al.*[2] in 1931 reported on the diagnosis of bone tumours by aspiration and in 1949 Siffert and Arkin reported their experience of trephine biopsies of bone.[3]

Since that time there have been many articles and much literature discussing the value of closed needle biopsy and stating the advantages and disadvantages of the procedure. Copious comparisons have been drawn with open needle biopsy techniques; as expected, articles written by proponents of closed biopsy of bone favoured this technique over open biopsies. There is little doubt that open biopsies of bone, particularly from deep structures, commonly subject a patient to a major procedure with extended recovery time.

The accuracy of diagnosis from needle biopsy compares favourably with that from open surgical biopsy,[4–6] particularly when multiple specimens are obtained from different parts of the bone to show normal and abnormal bone in continuity. The exact path of the needle is documented by taking a radiograph and relating this to the pathological specimen presented to the bone pathologist. Needle biopsy of bone is cost-effective and safe and, compared to surgery, it is relatively non-invasive. Unless total excision of a lesion is being contemplated as the definitive form of treatment irrespective of the biopsy findings, there is little justification to proceed with open biopsy as the primary diagnostic technique.

The decision to undertake a biopsy is taken in conjunction with the referring clinicians and according to the needs of an individual patient.

Percutaneous versus open biopsy

Proponents of the open biopsy method state that an open biopsy gives a more reliable diagnosis than needle biopsy, and that adequate tissue is obtained in approximately 90 per cent of patients.[7]

Surgeons maintain that they are able adequately to remove the entire biopsy scar and tract at the time of the definitive surgical procedure.[7] Fidler and Niers,[8] reporting on a single open transpedicular biopsy of a vertebral body, state that aggressive spinal tumours require a removal *en bloc* of the lesion and of any biopsy tract and haematoma caused by the biopsy. To achieve this, a frozen section is taken at the time of surgery, followed by an immediate hemivertebrectomy if malignancy is shown. They also claim that a slight directional error could damage the pedicle and lead to paravertebral or epidural contamination with tumour and that open transpedicular biopsy avoids these complications.

These statements are not backed by a scientific study. However, Mankin *et al.*,[9] in a series of 329 mostly open biopsies for musculoskeletal tumours, reported major biopsy wound problems in 17 per cent, errors in diagnosis in 18 per cent, and 10 per cent non-representive or technically poor biopsies. The optimum treatment plan had to be altered in 18 per cent because of problems associated with the biopsy. Waldron *et al.*[7] reported on 51 patients who had 55 biopsies and reported complications of wound infection, haematoma, inappropriate biopsy excision site and problems with the drain exit site. The accuracy of diagnosis was 92 per cent, and they consider that the major disadvantage of the open method was the risk of more widespread tumour seeding, along with the high complication rate of the procedure. Proponents of the percutaneous closed biopsy technique favour the approach because of the relative ease of performing the procedure, the ability to carry out the investigation under local anaesthetic, the allowance of immediate therapy without waiting for the incision to heal and the reduction in costs of the procedure.

Implantation of tumour in the local soft tissues, although reported occasionally,[5] cannot be accepted as a reason for avoiding biopsy if tissue needs to be obtained before undertaking logical treatment.

If careful planning is carried out using either the open or closed biopsy techniques, then the tract can be excised at the time of surgery. In any event, the majority of biopsies are undertaken for secondary deposits or disc infections and in these instances major surgery is usually irrelevant. The simplest procedure with the lowest complication rate should be selected.

Anaesthesia and analgesia

The majority of proponents of percutaneous bone biopsy have used local anaesthesia.[10–21] They have argued that general anaesthesia and its complications can be avoided and, except for the most fragile patients, most of the biopsies can be achieved on an outpatient[22] or, certainly, day bed basis. Some authors prefer their patients to be awake and conscious so that they can inform the surgeon or radiologist if they have any unusual symptoms during the biopsy.[23] Radiating pain might indicate nerve encroachment,[23] and would necessitate a change of direction of the biopsy needle, although this symptom cannot be relied upon for guidance as adequate local anaesthesia may anaesthetize the peripheral nerves.

A number of patients undergoing procedures under local anaesthetic, even after very adequate and extensive infiltration of the periosteum and surrounding tissues using lignocaine or an equivalent, still complain of a substantial amount of deep-seated bone pain when suction is applied during specimen removal.[24–26]

The proponents of general anaesthesia argue that a bone biopsy can be painful and harrowing. The condition of the patient can be carefully monitored by an anaesthetist, and general anaesthesia allows very adequate sampling to be made including extra specimens from different parts of the same or other bones.[27] There are occasions when, after local anaesthesia, a sample is difficult to obtain and a new approach needs to be made. If general anaesthesia is used, further biopsies are easily obtained, negating the need for reintroduction of local anaesthetic. Multiple bones can be sampled and only in very rare instances is there a problem with anaesthesia.

A sensible middle-ground approach would be that some biopsies can be undertaken using local anaesthesia and others require a general anaesthetic.[3]

The advantage of an anaesthetist to monitor the patient would seem to be obvious, allowing the radiologist to pay full attention to the often technically demanding procedure. A combination of local anaesthesia and neuroleptanalagesia is a very sensible alternative.

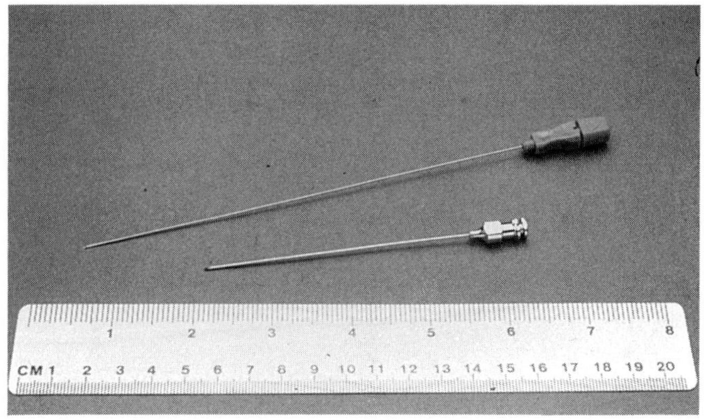

(a)

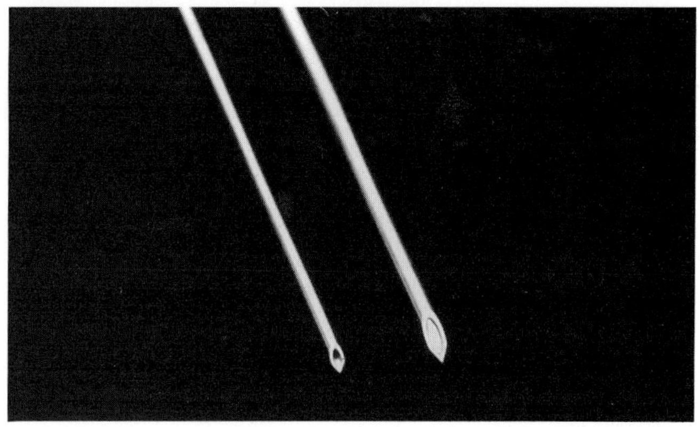

(b)

Fig. 5.1
(a) Spinal needles. (b) Tips of spinal needles.

BIOPSY NEEDLES

There are four major types of needles available for biopsy.

1 Spinal needles: These can be used in a variety of lengths and a variety of sizes, ranging from 16 to 22 G[1, 15, 28] (see Fig. 5.1). They are especially useful for infiltrating local anaesthetic[11,17] and for aspirating material for cytological or microbiological examination directly from soft tissue lesions, or the soft tissue component of bone tumours and areas of infection.[19] These needles can be inserted directly through the soft tissues or inserted through the trochar of a large needle.

   Some needles may provide a guide to a safe route for inserting a large needle,[12,29] especially for those radiologists and surgeons who are inexperienced in bone biopsy and who feel reassured by using this thin needle to guide the chosen route. Some authors describe the use of a modified bronchial brush passed through a 14 G needle to improve the diagnosis of cystic bone lesions.[30]

2 Cutting needles: The most commonly used is the Trucut needle, which can be hand fired or fired by using a triggering device.

   The specimen slot is capable of holding a small amount of soft tissue that has been sliced by the sharp outer cutting cannula (see Fig. 5.2). This type of needle can be used to obtain a core of soft tissue for histological examination and is ideally suited for those tumours that have a large, or even small, soft tissue component that may be accessible for histological examination.[21,26,31] The benefit of obtaining these types of specimen is that preparation for histology may be obtained more rapidly as bone specimens will need decalcification before sectioning.

3 Trephine cutting needles: These are large needles combining a central pointed obturator with an outer cannula. Once the needle, obturator and cannula have been positioned with the needle's tip adjacent to the lesion the obturator is withdrawn and the inner tooth cannula inserted into the bone.

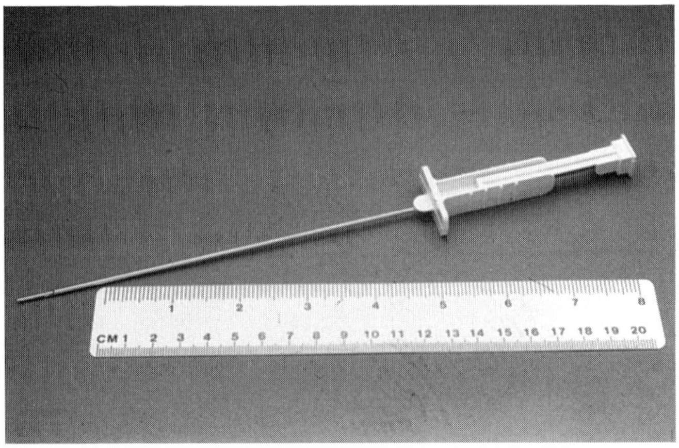

(a)

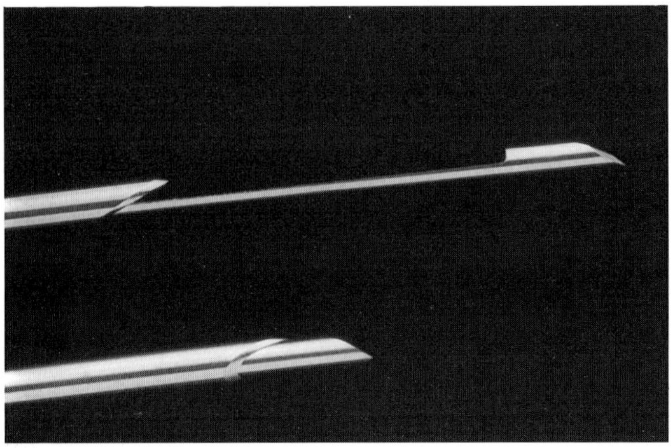

(b)

Fig. 5.2
(a) Trucut needle. (b) Specimen of Trucut needle open and closed.

Some needles, such as the Jamshidi needle,[10] come without a second outer cannula (Fig. 5.3(a)). There are many types of trephine needle, including those of Turkel and Bethel,[21,32] Ackermann,[21,33] Craig,[21,24] Radner,[34] Pais[12] and Fyfe.[11]

4 Drills: Further instruments using a principle similar to that of trephine needles have incorporated the use of an electric drill to facilitate the acquisition of some specimens.[35] The Corb needle is one such needle.[36]

The real breakthrough in biopsy is that specimen architecture is preserved when the trephine diameter is 2.0 mm or greater.[10,21] The factor which, in the authors' opinion, is absolutely critical is the use of a needle that tapers distally towards either a non-serrated or a serrated sharp tip and has a larger proximal shaft.[10,12,27] This type of needle is suitable for undertaking biopsies of both lytic and sclerotic lesions, but some needles will not cut through the very thick cortical bone of the femur, tibia and humerus. The tapering tip ensures that the cut specimen slips easily into the larger lumen of the body of the needle and prevents plugging, crushing and damage to the specimen inside the needle (Figs 5.3(b)–5.3(d). In this way long cores of tissue can be removed intact from the bone. They are expelled from the proximal end of the needle using a stylet. If a Jamshidi-type needle is used without an outer cannula then it is not possible to confirm that a core has been obtained until the needle has been withdrawn. Some authors have regarded this as being a problem in that if further cores of bone are required then the entire needle has to be inserted into the soft tissues with a potential risk of damaging further tissues. In the authors' experience, whether an outer cannula is being used or the entire needle is withdrawn does not alter the complication rate associated with biopsies. It is possible, if this theoretical problem is to be overcome, that an outer cannula can be made in order to achieve a single entry site into the bone being biopsied.

An excellent addition to the biopsy device is a metallic orthopaedic hammer,[16,34] which can be used to tap the needle into fairly sclerotic lesions. The use of straight, untapered needles will inevitably lead to crushed, distorted specimens being obtained.[14] The authors strongly recommend the use of tapered-tip needles.

Imaging guidance for biopsy

Accurate localization of a bone lesion and any soft tissue component must always be undertaken. A combination of plain radiographs, plain tomography, isotope bone scintigraphy, CT and MRI may be undertaken in order to obtain a complete picture of the bone lesion (see Fig. 5.4). In practice not all the above investigations need be used to study each bone abnormality. Newer image intensifier fluoroscopy machines have allowed very clear visualization of bone lesions and have marked an advance in the safety and accuracy of bone biopsy. Bone abnormalities are usually identified on plain films, but tomography may aid in the delineation of the area of abnormal bone. After the lesion has been visualized on conventional radiographs and possibly some form of axial tomography a diagnosis is often possible. Fluoroscopy will usually provide simple and easy guidance for the biopsy.[27] The needle can be seen dynamically entering the lesion.

Technique

Sterile technique is employed. The lesion having been localized with the most appropriate available imaging, 10–20 ml of 1% lignocaine is infiltrated along the biopsy track down to and under the periosteum using a 20 G

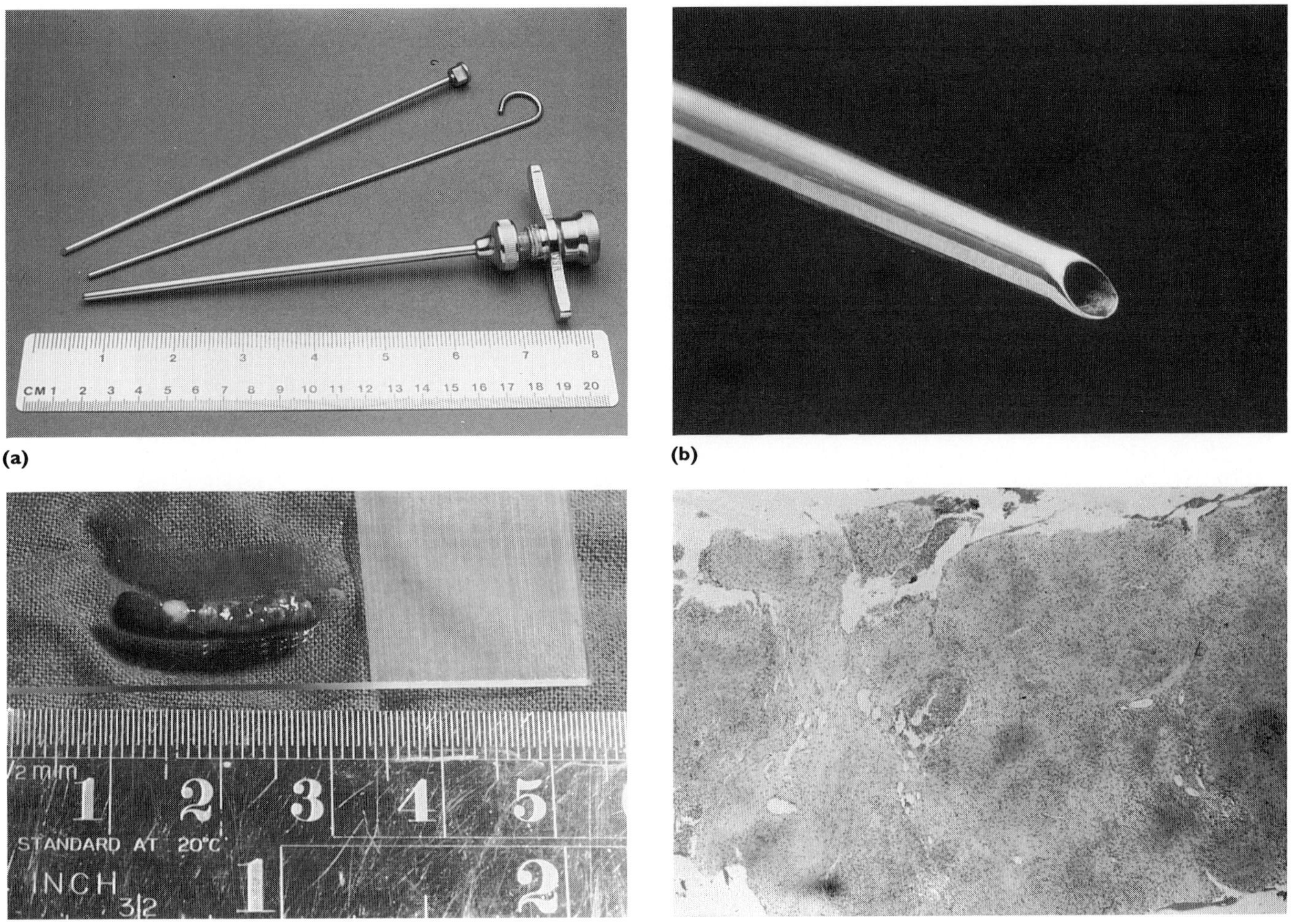

(a) (b) (c) (d)

Fig. 5.3
(a) Jamshidi needle. (b) Detail of bevelled end of Jamshidi needle. (c) Average-sized specimen core from Jamshidi needle. (d) Note the excellent size and quality of this Jamshidi needle biopsy histology specimen.

spinal needle. A small skin incision is made with a scalpel.

In the authors' opinion the most suitable commercially available needle is the Jamshidi needle. The obturator is locked into the cannula and the needle advanced towards the biopsy site. To minimize radiation to the operator, long forceps are used to hold the needle during screening. Once the needle tip has reached bone, its position is checked by fluoroscopy in two planes. If the position is satisfactory the obturator is pushed through the cortex and then removed. The cannula is firmly screwed into the lesion; a radiograph in one or two planes is then taken to document the biopsy site. Suction is applied to the proximal end of the needle using a 20-ml syringe while the needle is rocked back and forth to loosen the biopsy instrument and to detach the specimen. The needle is then removed while suction is maintained. The core is carefully ejected from the proximal end using a probe. The procedure is normally repeated twice.

Histology of bone will require several days for decalcification, followed by some time for the process of embedding in paraffin, sectioning and staining. In making the diagnosis of malignancy, blood clots obtained with the bone frequently yield a positive result.[14]

Smears for cytology may be stained immediately and examined using a microscope, or fixed with 95% ethanol and stained later. Other slides are air dried and stained later.

Single-plane fluoroscopy is adequate for many lesions, particularly the long bones, and simple rotation of a long bone can produce biplane assessment so that the depth of biopsy can be easily gauged. However, C-arm fluoroscopy is preferred, particularly for vertebral biopsies.[27]

By demonstrating unsuspected lesions in areas which were thought to be normal, bone scintigraphy helps in those patients with multiple bone lesions, in order to identify the most accessible site for percutaneous biopsy. Where clinical suspicion is high, fluoroscopically guided

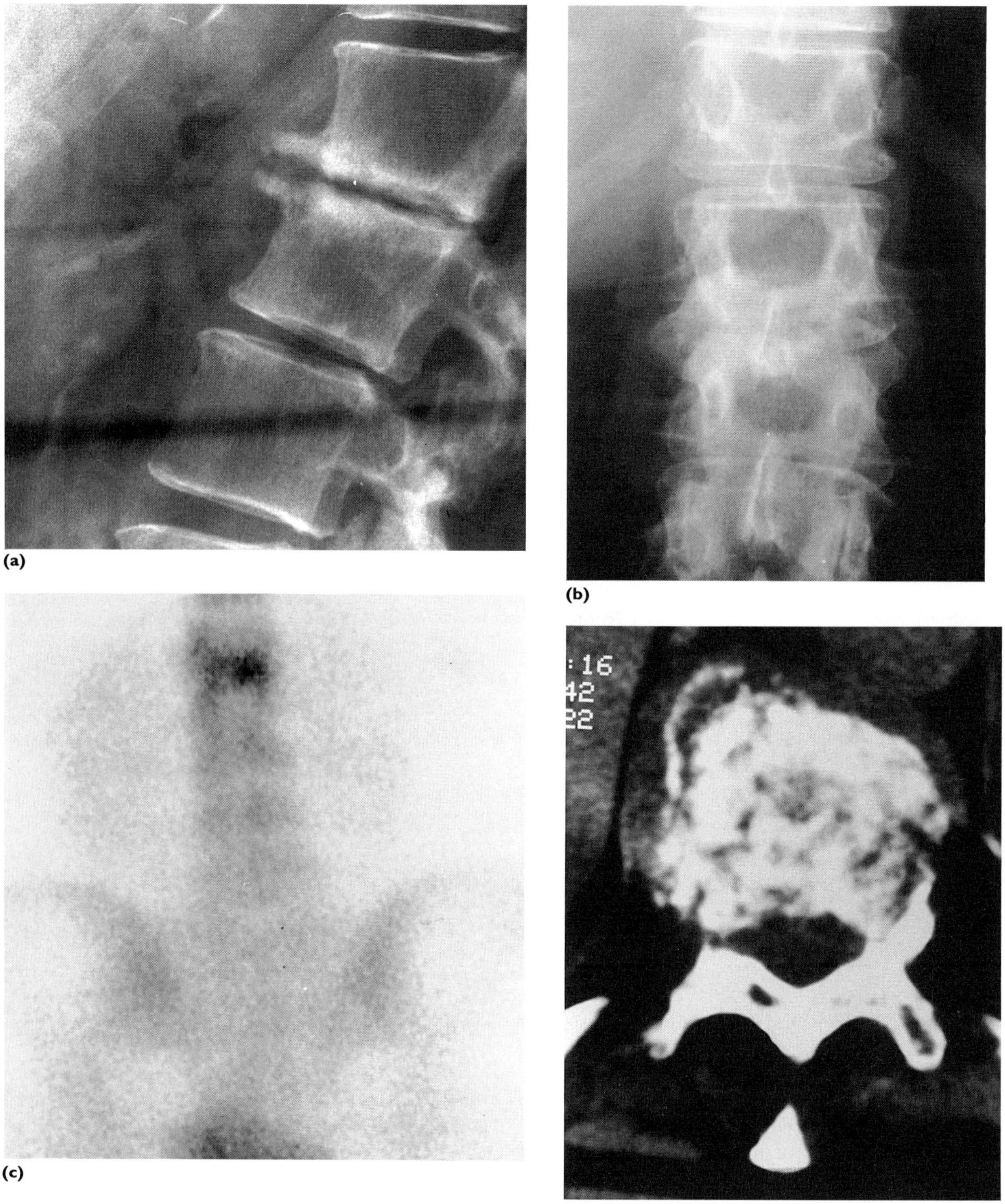

(a) (b) (c) (d)

Fig. 5.4
A 78-year-old woman presented with backache. (a), (b) Plain films showed reduction in disc space and sclerosis of adjacent vertebrae. (c) Bone scintigraphy showed this area to be of high activity, and (d) CT scan confirmed destruction at this level.

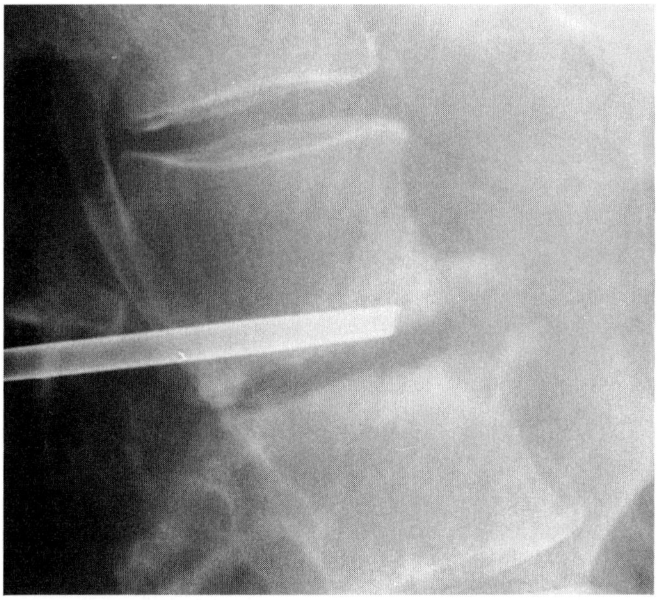

(e)

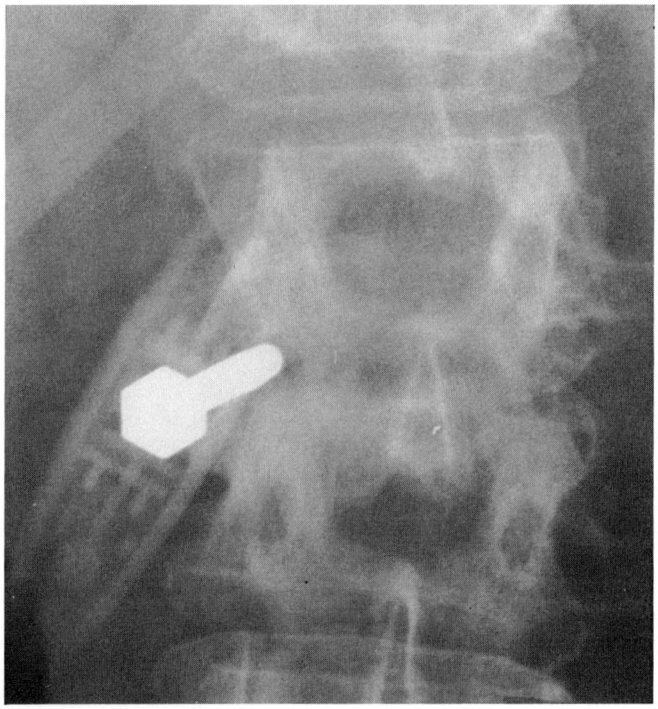

(f)

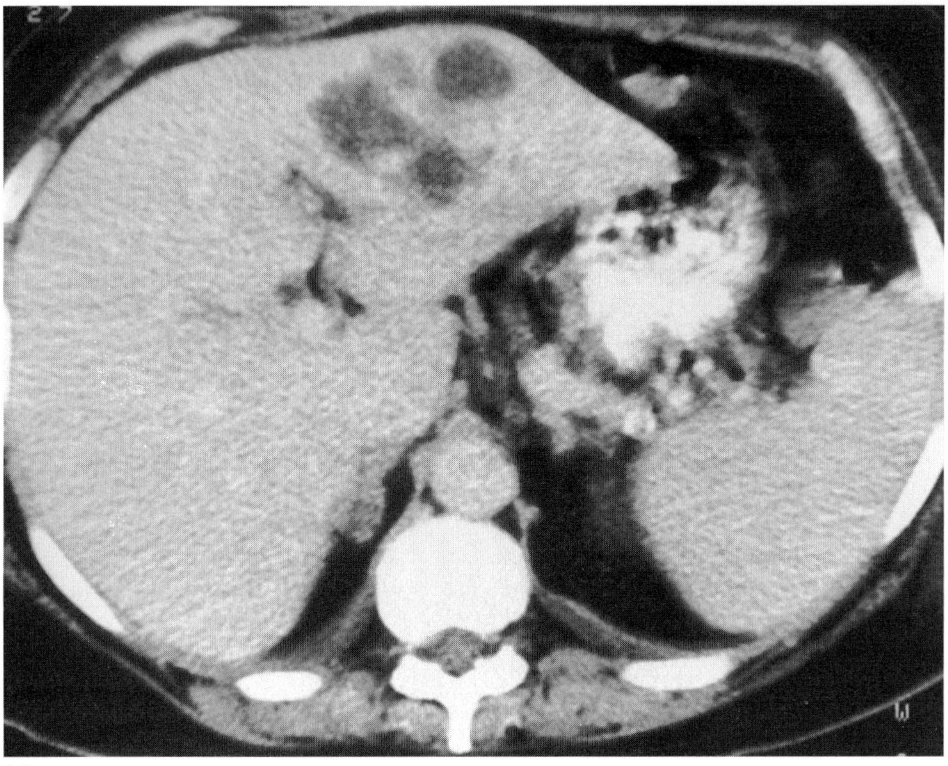

(g)

Fig. 5.4 *Continued*
(e), (f) Bone biopsy demonstrated change consistent with a discitis or osteomyelitis. (g) A low-density area in the liver was demonstrated. Liver biopsy grew *Streptococcus milleri*.

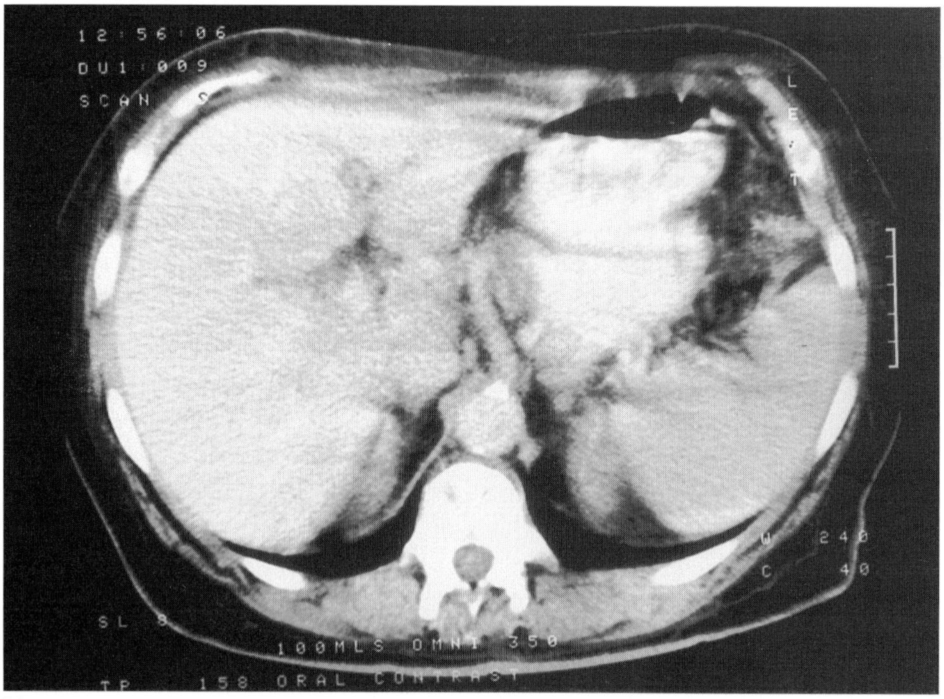

(h)

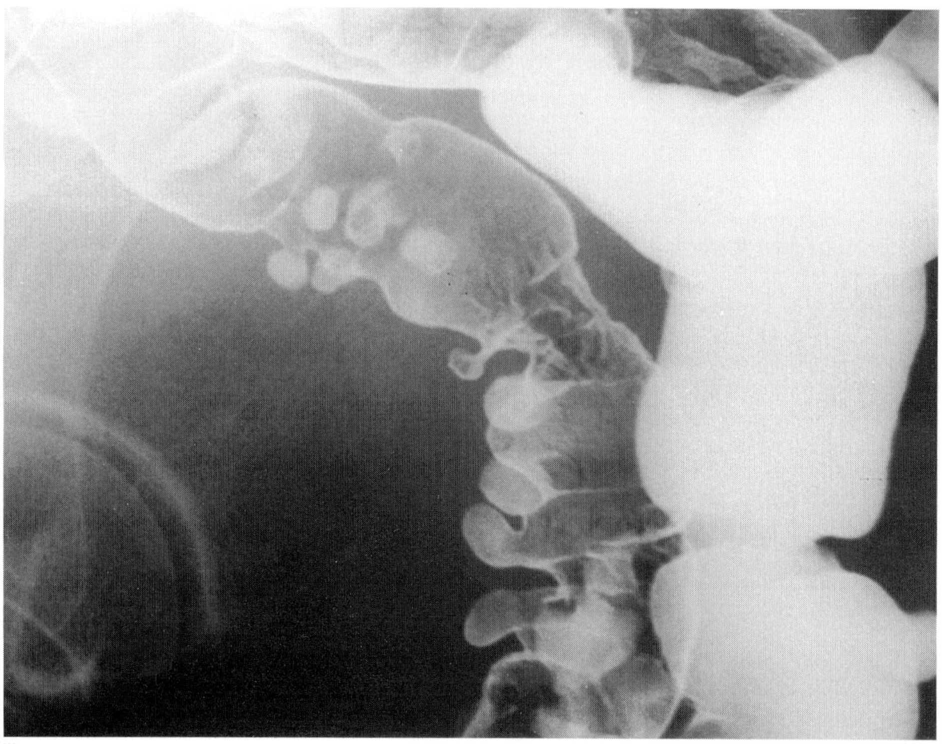

(i)

Fig. 5.4 *Continued*
(h) The liver lesion cleared and backache improved after treatment with appropriate antibiotics. (i) It is postulated that the focus of infection could have arisen from a sigmoid diverticulum.

biopsies of lesions localized by isotope studies give results similar to those for biopsies of lesions identified radiographically (over 80 per cent positive results).[37]

CT has a greater sensitivity than conventional radiographs in detecting metastatic lesions following abnormal findings from radionuclide bone scintigraphy,[38] and it may be used to localize abnormalities very accurately prior to biopsy. It is particularly useful in demonstrating the soft tissue component of bone lesions, which often contains the most aggressive region of tumour growth, or areas of infection (see Fig. 5.5). CT may also be used for needle guidance during the biopsy, though the thin slice

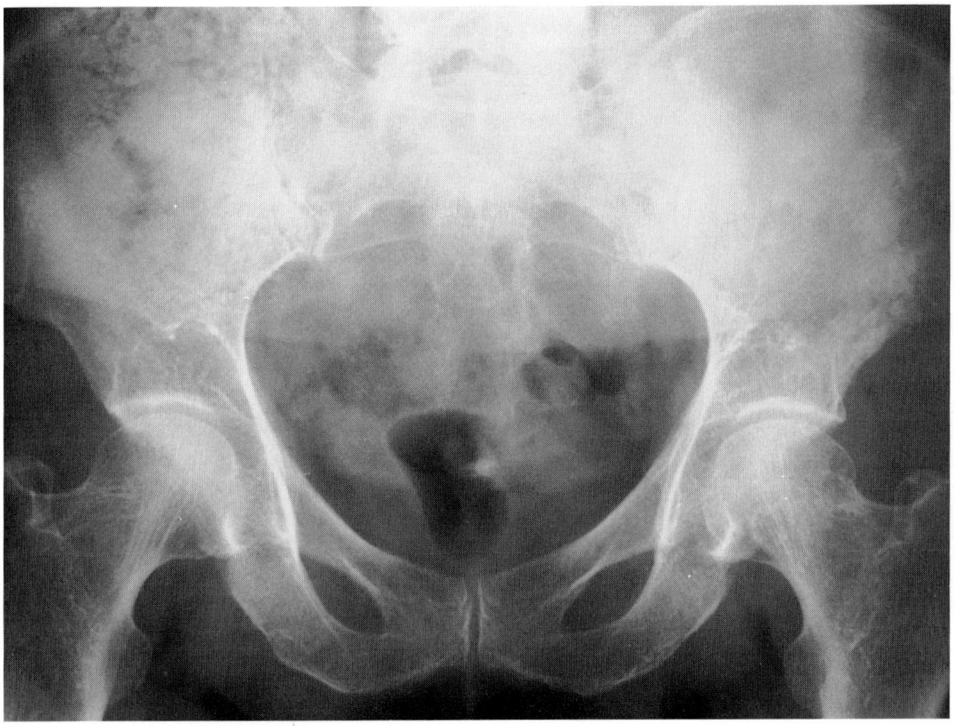

(a)

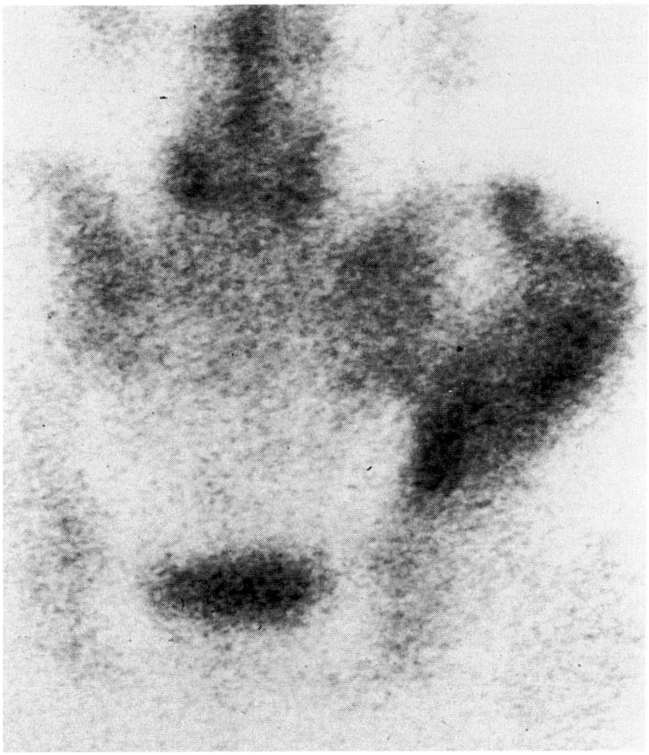

(b)

Fig. 5.5
A 75-year-old woman presented with hip pain. (a) The radiograph showed a lytic area over the left ilium (not easily visible). (b) Bone scintigraphy showed a high-activity area over the left ilium.

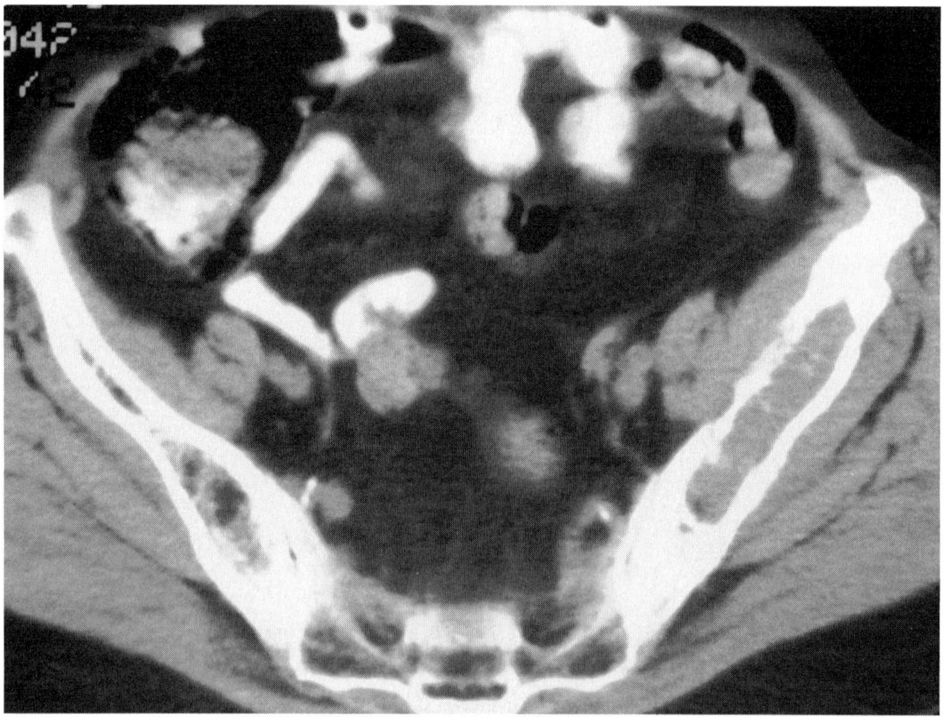

(c)

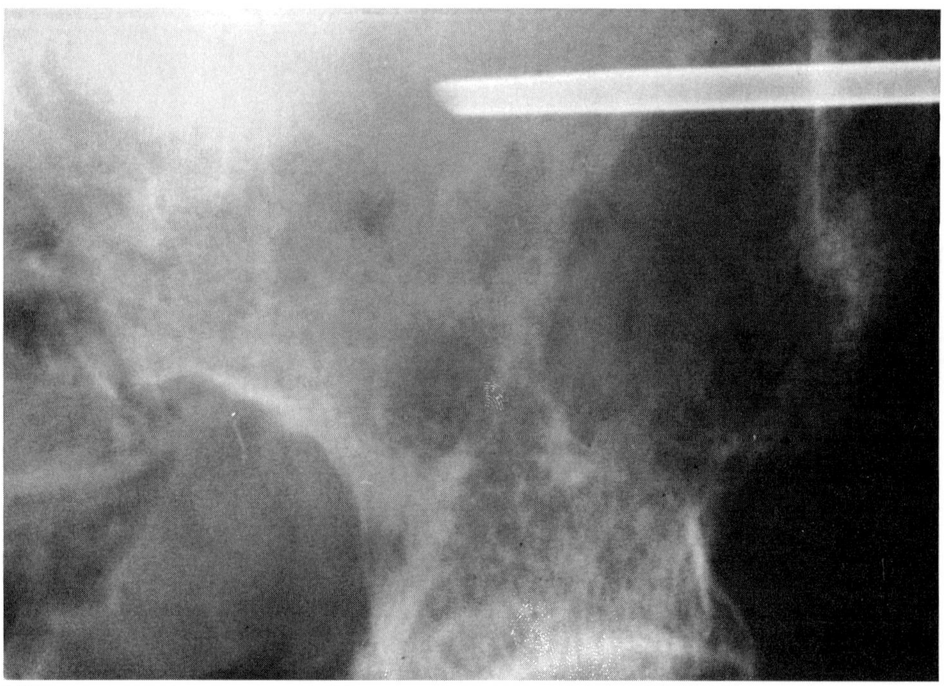

(d)

Fig. 5.5 *Continued*
(c) CT showed an obvious tumour with bony expansion and periosteal reaction, the tumour extending into the adjacent muscles. (d) Jamshidi biopsy demonstrated B-cell non-Hodgkin's lymphoma of low grade.

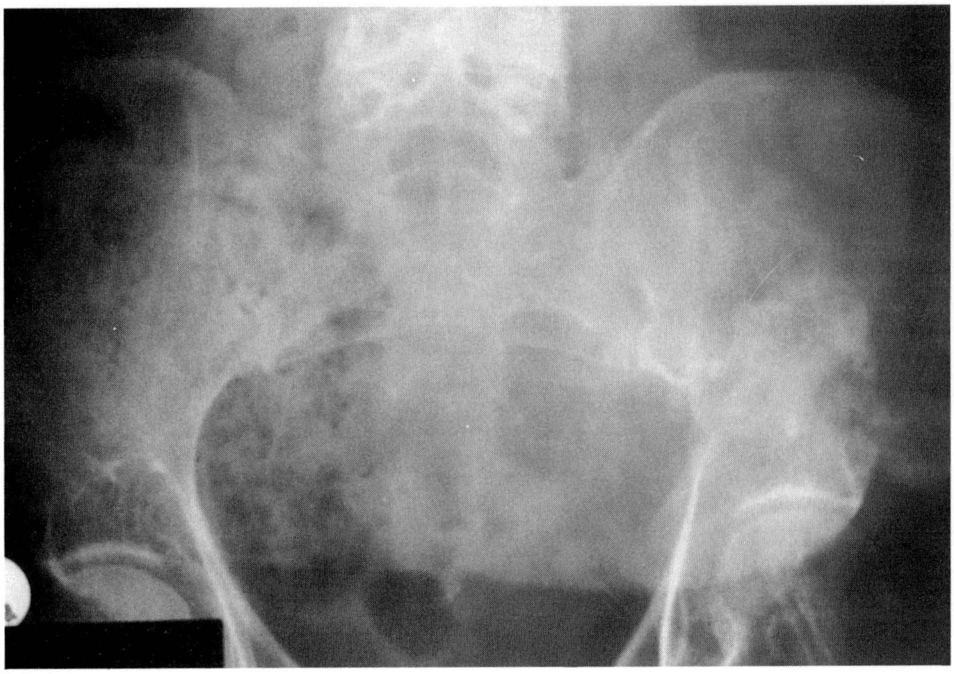

(e)

Fig. 5.5 *Continued*
(e) A post-radiation radiograph shows sclerosis consistent with healing.

thickness and inability to image dynamically during needle insertion are relative disadvantages.

However, there are a number of papers describing CT-guided percutaneous biopsy and this is a very acceptible method of obtaining tissue, particularly by those practitioners who are working almost exclusively with CT.[17–19,39]

INDICATIONS FOR BIOPSY

The commonest indication for closed bone biopsy is suspected metastatic bony lesions. The diagnosis of metastic disease can often be made with high confidence by its radiological features alone, but there are situations in which tissue sampling may alter diagnosis, prognosis and management.

A bone biopsy is indicated when:

1 There are apparent skeletal metastases without an obvious primary tumour. Histological diagnosis may shorten and simplify the search for a primary tumour and enable better management decisions to be made.
2 A known primary malignancy is associated with bony lesions atypical of metastases.
3 Apparent skeletal metastases are in disagreement with the clinical stage of the disease.
4 Radiographic images do not clarify a positive bone scan in a patient at risk of metastatic disease.

Another common indication for bone biopsy is suspected infection (see Figs 5.6(a, b) and 5.7(a)–(d)). The diagnosis of bone infection should be made using a combination of radiology, histology and microbiology because using a single criterion, such as positive bacteriology, may result in an artificially low diagnosis rate.[27,40]

In suspected eosinophilic granuloma,[41] giant cell tumour, [15,28] metabolic disease,[25,35] post-radiation bone necrosis and systemic conditions involving bone, percutaneous biopsy may be of great diagnostic value.[5,15,27,42–44]

Needle bone biopsy is excellent in diagnosing primary bone tumours.

Cartilage tumours and parosteal sarcomas tend to vary in aggression throughout the lesion and therefore multiple long samples should be taken from a number of sites in the tumour.

Radiologically benign skeletal neoplasms may be left without a biopsy but we have followed a policy of biopsy in order to fully document and classify most lesions as this will always allow optimum diagnosis and planning of therapy.

CONTRAINDICATIONS FOR BIOPSY

There are no absolute contraindications for a skeletal needle biopsy although we believe that an uncorrected bleeding diathesis can make the procedure hazardous. Although we have pursued a policy of undertaking biopsies of many benign bone lesions we have not taken

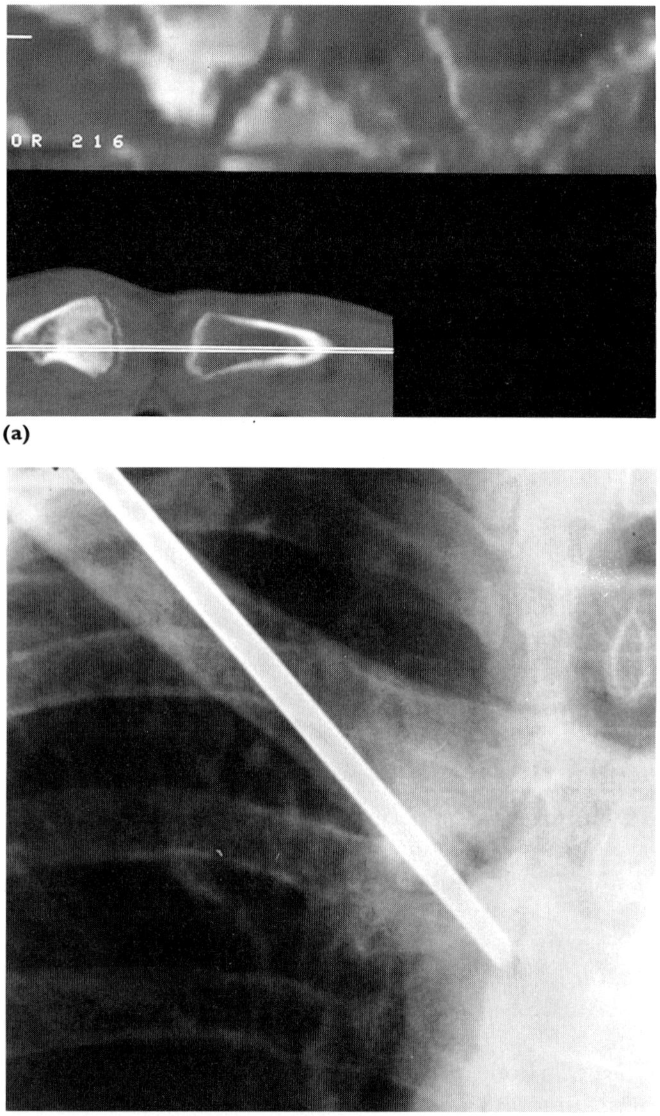

(a)

(b)

Fig. 5.6
(a), (b) The sclerotic medial end to the right clavicle was suspected to show an infective pathology, but bone biopsy revealed only erosive osteoarthritis.

biopsies of haemangiomas of the vertebral bodies and benign Pagetoid bone unless there is a suspicion of malignant change. As with any procedure, the risk of the biopsy must be calculated against other diagnostic methods and those of initiating therapy without a specific diagnosis.

Preparation of specimens

Three types of specimen are obtained, the first for histology, the second for cytology and the third for culture and sensitivity. Liaison with colleagues in the departments concerned is important in settling on a general policy of specimen handling and also in special cases where suspicion of different conditions would prompt the use of different fixatives, stains or culture media. The cytologist and microbiologist may be able to give immediate opinions on the samples obtained when this is likely to be advantageous. If the clinical and radiological findings point towards infection, more tissue can be directed for culture and sensitivity, including bone, pus and saline washings from the needle. Direct examination for aerobic, anaerobic bacteria and mycobacteria may be made.

Depending on the clinical status of the patient, the radiographs and the likely diagnosis, the bulk of the material is directed towards histology or microbiology. It is probably sensible to include cytology smears as well. In the authors' opinion it is prudent to direct some material for all three types of examination in almost all instances.

There is a significant rate of detection of malignant cells in blood clots, therefore these should also be sent for examination.[14]

Imprint or smear preparations may be made for cytology from the main histological specimen before it is fixed in formal saline. The specimen is pushed out of the proximal end of the Jamshidi needle on to a slide and a second slide is rubbed along the specimen to produce two imprints for cytology. One slide is air dried and the other is fixed.

Approach to biopsy at different sites

Extravertebral biopsy sites

Four factors determine the route of approach:

1 The route should avoid damage to viscera and neurovascular structures.

   When a biopsy of the lower limb is undertaken, an anterior approach is usually the best method, and to obtain a biopsy of the upper limb a posterior approach may be most sensible. For the iliac crest, sacrum and ischium a postero-lateral or postero-medial approach is planned in order to avoid the sciatic nerve. A biopsy of the pubis is taken from an anterior route, avoiding the femoral artery and nerve (Fig. 5.8).

2 The possibility of surgical resection should always be considered. If this is likely, then the needle track should be placed within the area of resection. This will allow excision of the needle track and reduce any possibility of tumour seeding along the track.

3 The shortest possible safe, perpendicular path from the skin to the biopsy site should be chosen.

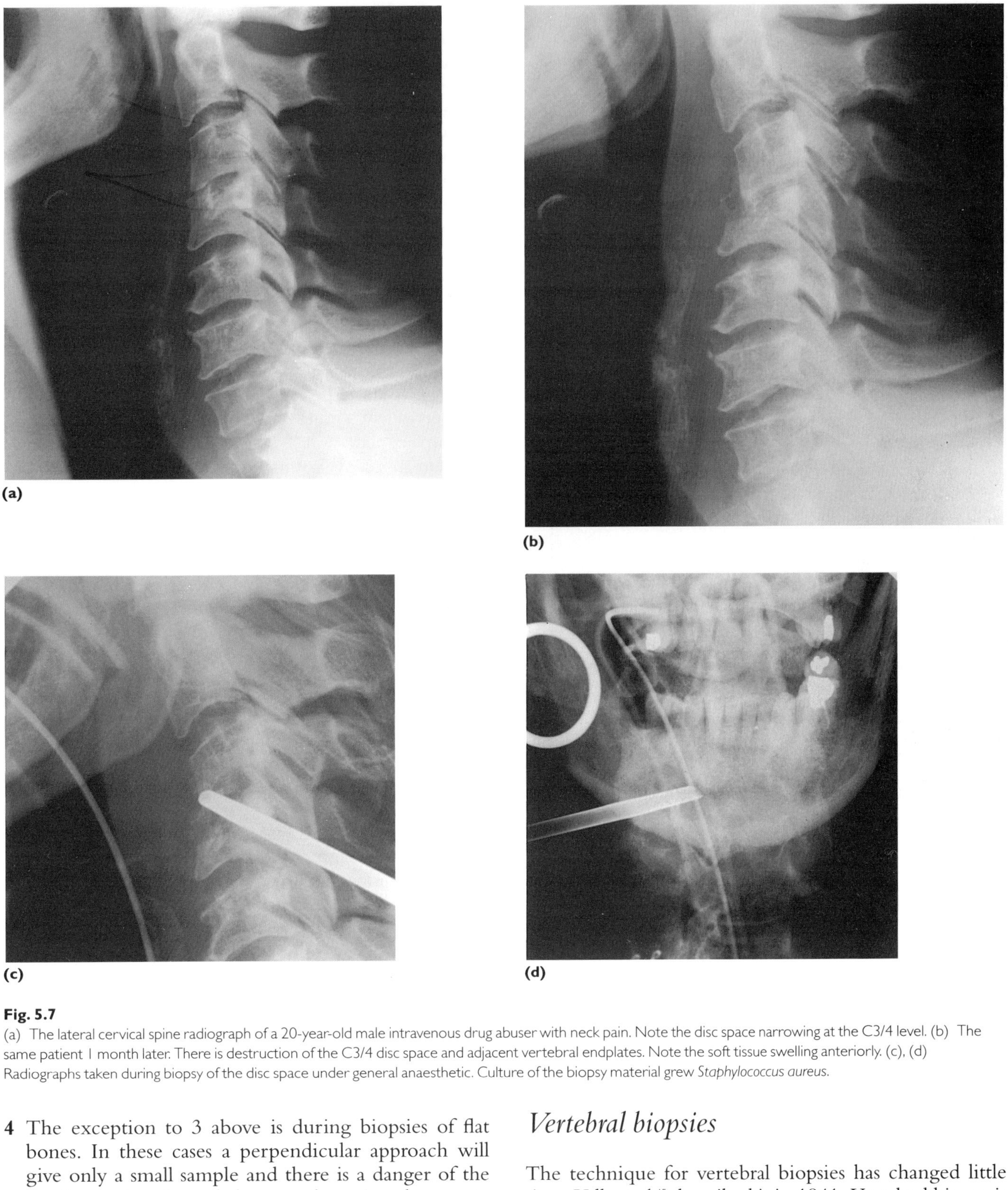

Fig. 5.7
(a) The lateral cervical spine radiograph of a 20-year-old male intravenous drug abuser with neck pain. Note the disc space narrowing at the C3/4 level. (b) The same patient 1 month later. There is destruction of the C3/4 disc space and adjacent vertebral endplates. Note the soft tissue swelling anteriorly. (c), (d) Radiographs taken during biopsy of the disc space under general anaesthetic. Culture of the biopsy material grew *Staphylococcus aureus*.

4 The exception to 3 above is during biopsies of flat bones. In these cases a perpendicular approach will give only a small sample and there is a danger of the needle passing through bone and causing damage to the deep structures, i.e. the pleura and lung posterior to the ribs. For these bones an oblique approach with an angle varying from 50 to 80° ensures that a longer, more representative specimen will be obtained.

Vertebral biopsies

The technique for vertebral biopsies has changed little since Valls *et al.*[45] described it in 1941. Vertebral biopsy is not a difficult technique, but, because of the proximity of major neural and vascular structures, extreme care must be taken and the route of approach needs to be carefully thought out.

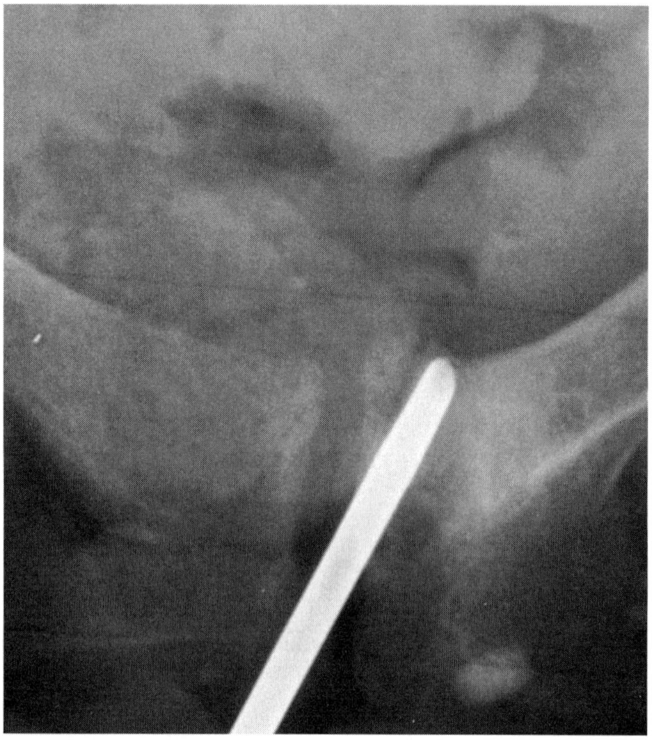

Fig. 5.8
Anterior needle route to pubis to sample an area of sclerosis which histology showed to be healing osteoporosis.

Biplane fluoroscopic control or a C-arm is desirable for biopsy guidance. Different regions of the spine should be considered separately.

1 The first two cervical vertebrae are approached anteriorly through the mouth with the patient under general anaesthesia[16] (see Fig 5.9(a)). A spinal needle is introduced via a mid-line approach into the vertebral lesion under careful fluoroscopic control and aspiration can be carried out. In these instances it is considered essential to have multiple methods of viewing the lesion so that other routes of approach can be considered. It is possible to do a postero-lateral biopsy of a lateral mass or a lateral biopsy of a vertebral body.
2 The remainder of the cervical spine can be biopsied using a Jamshidi needle. The patient is placed supine and the carotid vessels carefully palpated. The entry site is chosen at the level of the vertebral body to be biopsied, usually posterior to the posterior margin of the sternocleidomastoid muscle and the palpable carotid artery. Under biplane fluoroscopy the needle is advanced, keeping posterior to the main neurovascular bundle and passing along the anterior border of the transverse process of the vertebral body (Fig. 5.9(b)).

   Screening in the AP and lateral views will show the exact position of the needle, and radiographs should be taken to document the needle site (see Fig. 5.7(c)–(d)). This will assist the histologists and will also be useful if, on a rare occasion, a second biopsy needs to be undertaken.
3 For the thoracic spine the approach for biopsy is posterior. The patient lies either prone or lateral, but we favour the prone approach. A skin entry site is chosen which is at the level of the body on which the biopsy is to be carried out, over the lateral edge of the pedicle of the vertebra. If the lesion lies on one side of the vertebra, that side is chosen for the approach. If the abnormality is on both sides of the vertebra equally then the right side is chosen to avoid any risk of damaging the aorta. The angle of entry is between 5 and 30° (see Fig. 5.10), but probably, most sensibly, about 10° from the perpendicular. This should carry the needle through the transverse process, sometimes the rib edge, through the pedicle and into the vertebral body, avoiding the spinal canal contents and the pleural reflection. We prefer this technique to a more oblique approach.
4 A similar point of entry may be used for lumbar and thoracic spinal biopsies. A more lateral approach may be used. A skin entry site 6–8 cm from the mid-line, just lateral to the edge of quadratus lumborum muscle and between the lower ribs and iliac crest, is chosen. The angle of entry of the needle is approximately 45° (Fig. 5.11). Again, the right side is preferred and the route of approach is chosen to avoid the transverse process. If the transverse process is encountered then slight angulation can achieve a satisfactory biopsy route.

A number of reports have stated that cartilaginous tumours cannot be graded by small samples obtained surgically or through a needle biopsy.[22] In the authors' experience, if biopsies are undertaken under general anaesthesia then large, multiple samples will be adequate for histologic grading. The use of tapered-end needles of adequate size will produce excellent biopsy specimens for secondary or primary skeletal neoplasms or infection. Some authors have stated that needle biopsy is not adequate for definitive subclassification of a sarcoma.[22] Our experience is completely opposite to this. Correct needle choice and a range of biopsy samples allow very adequate histological grading. Parosteal sarcomas can also be biopsied in the above manner.[27,42]

Complications

Complications are relatively few. If local anaesthetic is used, pain is the most frequent side-effect and is usually caused by inadequate local anaesthesia.

Negative pressure and manipulation within the medullary canal may result in severe deep-seated pain. Neurological damage, often paraesthesia but even

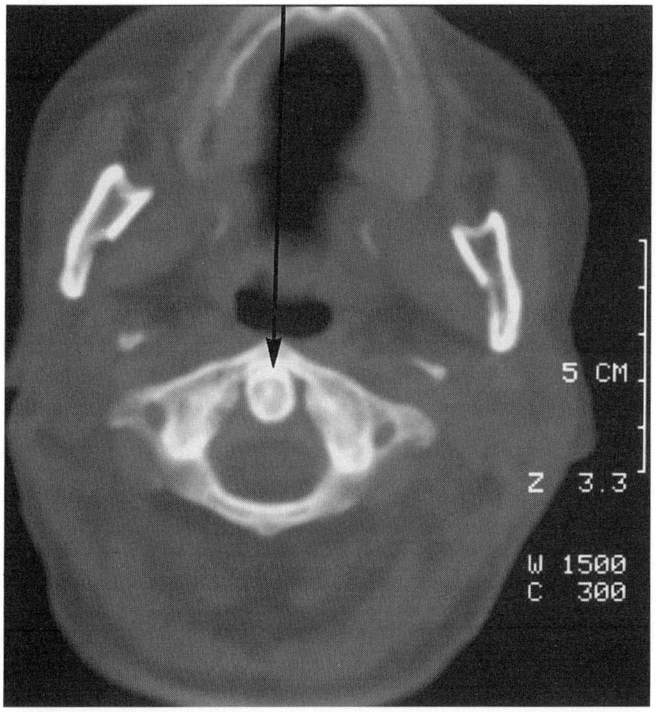

(a)

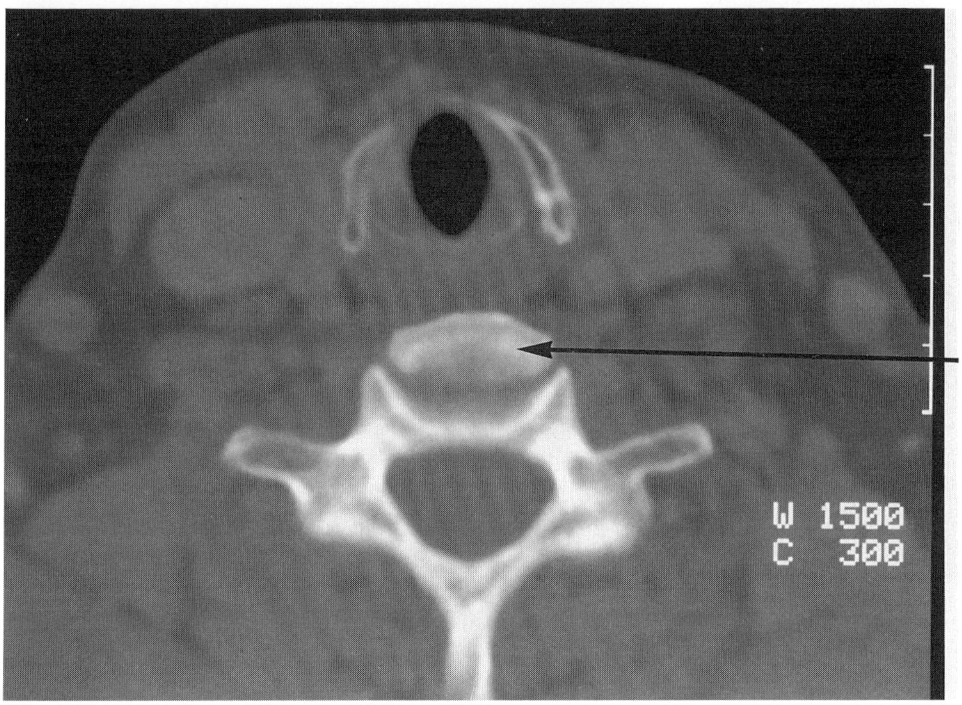

(b)

Fig. 5.9
(a) Open mouth position for needle biopsy of C1/2. (b) Lateral approach for biopsy of lower cervical spine.

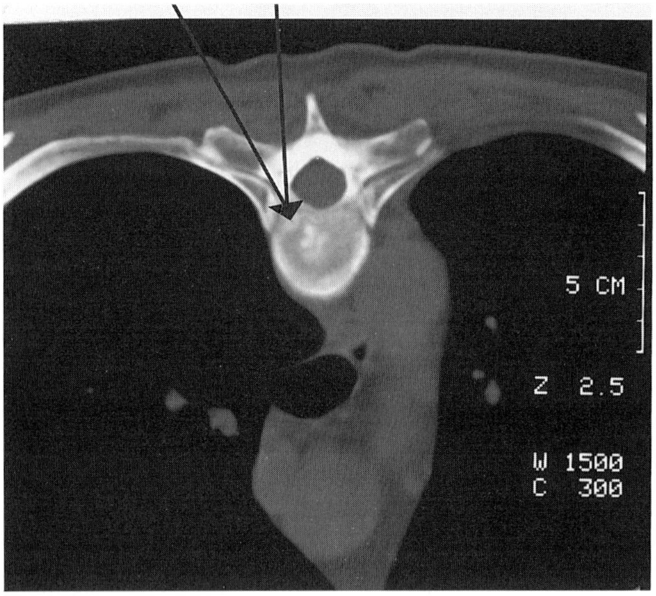

Fig. 5.10
Posterior needle approach for thoracic spine biopsy 5–30° from the perpendicular.

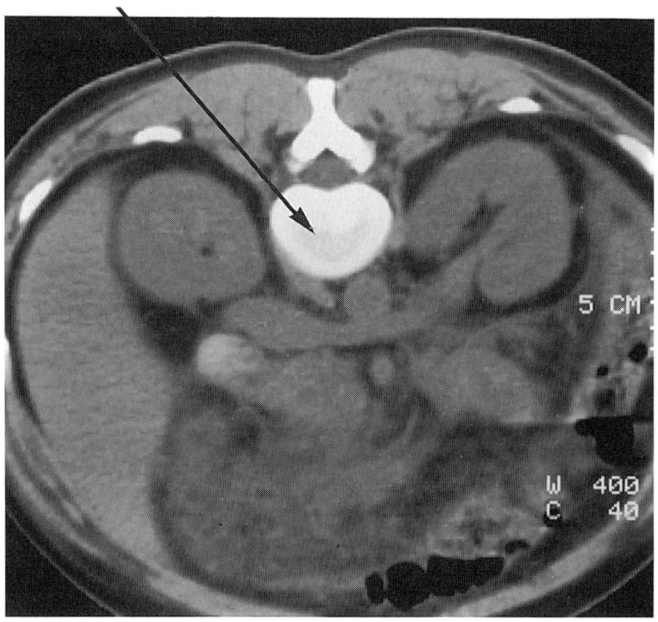

Fig. 5.11
Needle approach for lumbar spine biopsy 45° from the perpendicular.

quadraparesis, has been described following biopsy of the cervical spine.[46] Pneumothorax may occur in biopsies of the thoracic spine.[11,47] Haemorrhage can occur if large vessels are lacerated and tuberculosis may seed along sinus tracts secondary to the biopsy.[11,47]

Spread of tumour along the biopsy tract is so rare that, although occasionally reported,[5] it cannot be accepted as a reason for avoiding a biopsy procedure (see Fig. 5.12).

After-care of the patient

If local anaesthesia has been used then close nursing is not required. If, however, patients have undergone general anaesthesia then special nursing is required until the patient recovers fully from the anaesthetic. A chest radiograph is obtained routinely following thoracic vertebral biopsies.

Clinical aspects and conclusion

Needle biopsy of bone has various advantages over open biopsy. It is a safe procedure and allows repetition of biopsies if there is a failure to gain adequate diagnostic material.

Follow-up material can also be obtained following treatment and the biopsy needle can easily reach areas which are otherwise only accessible by major surgery. The small needle tract allows immediate treatment by radiotherapy without waiting for healing of an open biopsy wound and there is no scar to interfere with subsequent surgical treatment. The possibility of fracture following a biopsy is low.

As an operator gains more experience with bone biopsy so larger needles are used, reducing the possibility of sampling errors. The use of adequate anaesthesia, be it local, general or a combination of the two, will allow multiple specimens to be obtained, providing many representative samples from all parts of the bone lesion.

We do not believe that cartilaginous tumours or parosteal sarcomas present any degree of difficulty of biopsy and interpretation if sufficient samples are taken. Cystic and necrotic lesions do not produce problems of diagnosis if biopsies are taken at the periphery of lesions. On the other hand, the diagnosis of a solitary bone cyst is difficult histologically and is made on clinical and radiographic findings.[43]

The major difficulty is to take a biopsy through very thick cortical bone such as the femur, tibia and humerus, and for this reason new biopsy needles and handles have been developed. The use of a strengthened biopsy tip, mechanical rotating devices and a metallic hammer will often ensure that an adequate specimen is obtained.

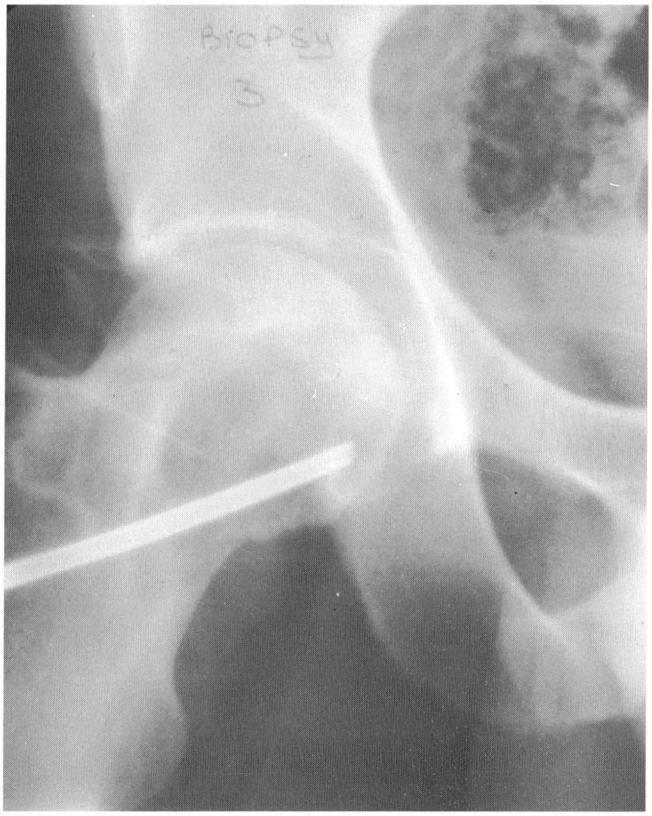

(a)

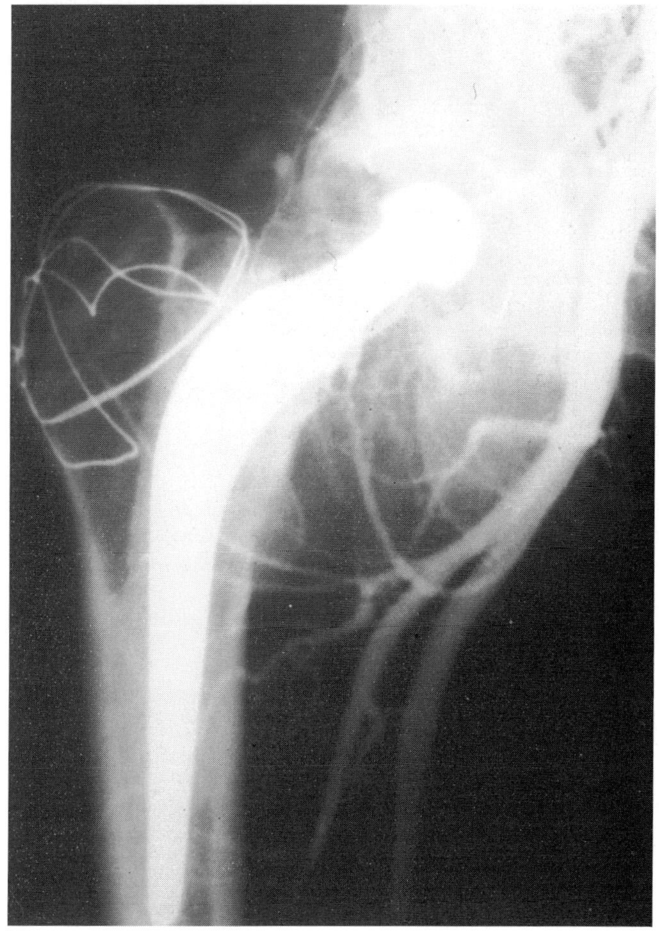

(b)

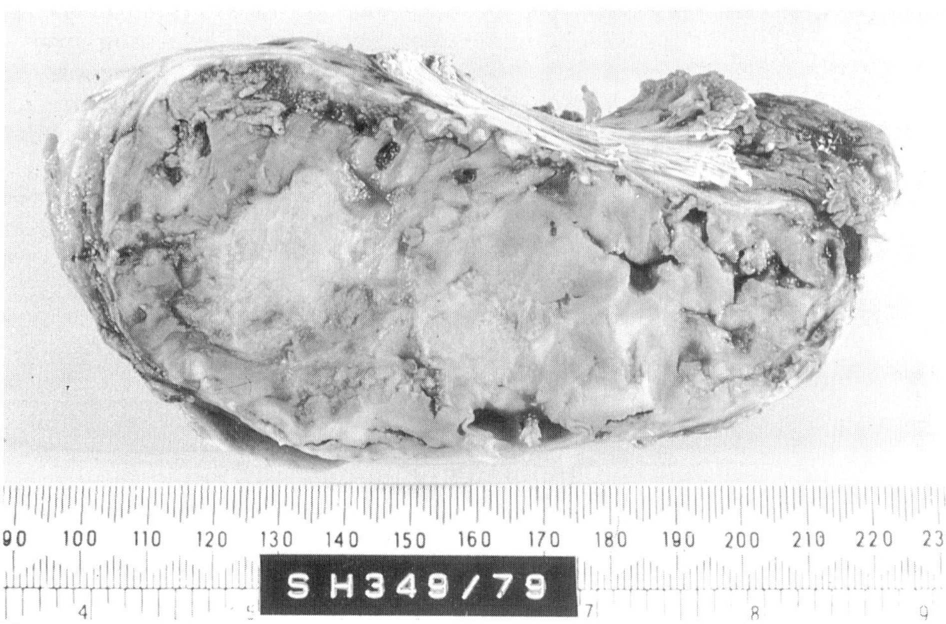

(c)

Fig. 5.12
(a) Giant cell tumour of femoral head extending to the articular margin. The tumour was excised and the hip replaced with a prosthesis. One year later the patient complained of a swelling in the groin. (b) The arteriogram shows medial displacement of the femoral artery. (c) A large recurrent tumour mass was resected with part of the femoral artery. This was called a needle tract recurrence by the surgeon although we believe this could be regarded as recurrence following surgery.

Authors	Numbers	Selection	Needle type	Areas biopsied	Positive diagnosis (%)	Insufficient material (%)	Non-conclusive, incorrect or false negative diagnosis (%)	Complications
Schajowicz and Derqui (1968)[43]	4050 bx 941 vertebral	Bone and soft tissue lesions biopsied over 25 years	2 mm or 4mm sharp needle	3717 bone, 295 joint, 39 soft tissue	76	n.r.	24	None serious reported
Ottolenghi (1969)[48]	1078	Spinal lesions biopsied over 28 years	2 mm diameter needle	28 thoracic, 78 cervical, 972 lumbar	73	n.r.	27	None serious reported
Deeley (1972)[6]	100	Radiographic lesions considered malignant	Trephine or pneumatic drill	Various	96	None	4	None
Gladstein and Grantham (1974)[49]	188	Radiographic lesions	Craig	Various	95	3% of patients unable to tolerate biopsy under LA	2	2 pneumothorax, 1 pneumonia
Debnam and Staple (1975)[47]	73	Radiographic and radioisotope lesions	Craig	Various	81	7	12	1 sinus track, 1 pneumothorax
Armstrong *et al.* (1978)[50]	19	Suspected disc space infection	Ackerman	Vertebral discs	68	None	32	1 abscess, 1 pneumonia
de Santos *et al.* (1979)[44]	34	Suspected primary bone tumours	Various	Various	88	7	5	None
de Santos and Zornoza (1980)[51]	379	Radiographic lesions	Various	Various	82	Biopsy repeated	18	None serious reported
Shaltot *et al.* (1982)[5]	50	Radiographic lesions	Jamshidi	Various	88	n.r.	12	1 parasthesia, 1 recurrence
Tehranzadeh *et al.* (1983)[20]	120	Radiographic lesions	Craig, Ackermann	Various	72	14	14	None
Fyfe *et al* (1983)[11]	100	Radiographic lesions	Harlow Wood	Vertebral	70	n.r.	30	15 neuro-logical, 6 infective, 5 pulmonary
Ayala and Zornoza (1983)[42]	222	Suspected primary bone tumours	n.r.	Various	79	16	6	None serious reported
Stoker and Kissin (1985)[4]	135	Radiographic lesions	Jamshidi, Ackermann	Vertebral	89	7	5	2 bleeds, 1 urinary retention
Laredo and Bard (1986)[52]	41	Thoracic lesions	Trephine with guidewire	Thoracic spine	76	0	13	None serious reported
Murphy *et al.* (1981)[21]	169 biopsies in 164 patients, 94 vertebral	Radiographic lesions	Various	94 vertebral, 75 various	94	2	4	2 pneumothorax, 1 requiring chest drain
Cotty *et al.* (1988)[40]	30 patients	Suspected vertebral osteo-myelitis	Various	Spinal and iliac bones	90 Histology + bacteriology	10		None
Brugieres *et al.* (1991)[39]	89 biopsies	Radiographic lesions	Trephine with guidewire	Vertebral	97	0	3	None
Odendaal and Lemmer (1991)[53]	55 patients	Radiographic vertebral body lesions	Jamshidi + trephine	Vertebral	44 true positive only	8	53	1 internal jugular vein puncture, 1 TB meningitis
Pais *et al.* (1984)[12]	11	Bone lesions	Pais	Bone, multiple sites	82	n.r.	18	n.r.
Hewes *et al.* (1983)[14]	47	Bone lesions	Variety	Various	94% osseous blood clot + 72% bone core	4% blood clot + smears 15% bone core	3% blood clot + smears 7% bone core	n.r.

Authors	Numbers	Selection	Needle type	Areas biopsied	Positive diagnosis (%)	Insufficient material (%)	Non-conclusive, incorrect or false negative diagnosis (%)	Complications
Vetrani *et al.* (1990)[28]	9	Radio-graphically, giant cell tumour	22G spinal needle	Various	100	10 repeated aspiration	0	None serious
Dollahite *et al.* (1989)[15]	57 patients 69 biopsies	Radiographic primary neoplasm bone	20 or 18G Craig needle	Various	57	33% insufficient material or incorrect diagnosis 10% no definite diagnosis		
Brugieres *et al.* (1990)[18]	27	CT-detached vertebral neoplasm	Trephine	Thoracic spine	89	7	4% false negative	None
Gatenby *et al.* (1984)[19]	12	Lytic lesion	22G needle	Various	100	0	0	None

Complication rate per patient (where specified) = 42/7032 patients = 0.6 per cent. Giving equal weight to all studies, positive diagnosis = 84 per cent (true +ve, true-ve).
Key: n.r. = not reported. n.a. = not applicable.

Table 5.1 *Summary of results of bone biopsy following a review of the literature*

Authors	Numbers	Selection	Needle type	Areas biopsied	Positive diagnosis (%)	Insufficient material	Non-conclusive, incorrect or false negative diagnosis (%)	Complications
Ackermann *et al.* 1956[33]	233 biopsies in 150 patients	Known primary; many had no radiographic lesion	Ackermann	Vertebral, and various	11	n.r.	n.r.	None serious reported
Dombernowksy *et al.* (1974)[34]	522 biopsies 200 patients	Leukaemia lymphoma	Radner	Bone marrow	n.a.	1.1	n.r.	None serious
Jamshidi and Swain (1972)[10]	150 biopsies in 131 patients	Haematologic, non-haema-tologic	Jamshidi	Posterior iliac crest	16	0	n.a.	n.r.
Lawrence *et al.* (1984)[13]	126	Small cell carcinoma of the lung	Aspirate and biopsy	Posterior superior iliac crest	24	n.a.	n.a.	n.r.
Dahlberg *et al.* (1989)[25]	19 biopsies in 18 patients	renal osteo-dystrophy	Jamshidi 11 Gauge	Posterior or anterior iliac crest	100%	0	n.r.	None

Key; see Table 5.1

Table 5.2 *Biopsy of bone where no focal radiographic lesion is present*

CLINICAL VALUE

Many series in the literature do not classify their biopsy results into true positive, true negative, false positive and false negative categories, and it is therefore not always possible to compare results.

Another inconsistency is the division between true positive and inconclusive results. Some studies have counted histological evidence of pyogenic infection as a true positive even when an organism has not been isolated. Others have considered such a result as inconclusive. We have attempted to summarize the biopsy results reported in the literature (Tables 5.1 and 5.2).

In order to optimize the rate of identification of organisms when infection is suspected, pieces of bone may be crushed and cultured, in addition to the aspiration samples. However, in our experience this has not led to a greater number of cultures being positive.

The overall complication rate for percutaneous biopsy is extremely low, and complications are mostly minor (Table 5.1). Serious complications are rare, although there have been isolated cases of transient quadraparesis,[46] paraplegia and death[54] following a needle biopsy.

REFERENCES

1 MARTIN HE, ELLIS EB (1930). Biopsy by needle puncture and aspiration. *Annals of Surgery* **92**: 169–81.
2 COLEY BL, SHARP GS, ELLIS EB (1931). Diagnosis of bone tumours by aspiration. *American Journal of Surgery* **13**: 215–24.
3 SIFFERT RS, ARKIN AM (1949). Trephine biopsy of bone with special reference to the lumbar vertebral bodies. *Journal of Bone and Joint Surgery* **31A**: 146–9.
4 STOKER DJ, KISSIN CK (1985). Percutaneous vertebral biopsy: A review of 135 cases. *Clinical Radiology* **36**: 569–77.
5 SHALTOT A, MICHELL PA, BETTS JA, DARBY AJ, GISHEN P (1982). Jamshidi needle biopsy of bone lesions. *Clinical Radiology* **33**: 193–6.
6 DEELEY TJ (1972). The drill biopsy of bone lesions. *Clinical Radiology* **23**: 536–40.
7 WALDRON B, MORAN RM, HURSON BJ (1990). Biopsy problems in patients with musculoskeletal tumours. *Irish Journal of Medical Science* **159**: 246–8.
8 FIDLER MW, NIERS BBAM (1990). Open transpedicular biopsy of the vertebral body. *Journal of Bone and Joint Surgery (British volume)* **72B**: 884–5.
9 MANKIN HJ, LANGE TA, SPANIER SS (1982). The hazards of biopsy in patients with malignant primary bone and soft tissue tumours. *Journal of Bone and Joint Surgery* **64A**: 1121–7.
10 JAMSHIDI K, SWAIN WR (1972). Bone marrow biopsy with unaltered architecture: A new biopsy device. *Journal of Clinical Medicine* **77**(2): 135–342.
11 FYFE IS, HENRY AP, MULHOLLAND RC (1983). Closed vertebral biopsy. *Journal of Bone and Joint Surgery* **65B**: 140–3.
12 PAIS MJ, LIGHTFOOTE KB, BURNETT K, EDWARDS KC, CREMER M, ORTIZ GT (1984). Trephine bone biopsy system: A refined needle for radiologists. *Radiology* **153**: 253–4.
13 LAWRENCE JB, ELEFF M, BEHM FG, JOHNSTON CL Jr (1984). Bone marrow examination in small cell carcinoma of the lung. *Cancer* **53**: 2188–90.
14 HEWES RC, VIGORITA VJ, FREIBERGER RH (1983). Percutaneous bone biopsy: The importance of aspirated osseous blood. *Radiology* **148**: 69–72.
15 DOLLAHITE HA, TATUM L, MOINUDDIN SM, CARNESALE PG (1989). Aspiration biopsy of primary neoplasm of bone. *Journal of Bone and Joint Surgery* **71A**: 1166–9.
16 PATIL AA (1989). Transoral stereotactic biopsy of the second cervical vertebral body: Case report with technical note. *Neurosurgery* **15**: 999–1001.
17 RENFREW DL, WHITTEN CG, WIESE JA, EL-KHOURY GY, Harris KG (1991). CT-guided percutaneous transpedicular biopsy of the spine. *Radiology* **180**: 574–6.
18 BRUGIERES P, GASTON A, HERAN F, VOISIN MC, MARSAULT C (1990). Percutaneous biopsies of the thoracic spine under CT guidance: Trans-costovertebral approach. *Journal of Computer Assisted Tomography* **14**(3): 446–8.
19 GATENBY RA, MULHERN CB, MOLDOFSKY PJ (1984). Computed tomography guided thin needle biopsy of small lytic bone lesions. *Skeletal Radiology* **11**: 289–91.
20 TEHRANZADEH J, FREIBERGER RH, GHELMAN B (1983). Closed skeletal needle biopsy. Review of 120 cases. *AJR* **140**: 113–15.
21 MURPHY WA, DESTOUET JM, GILULA LA (1981). Percutaneous skeletal biopsy: A procedure for radiologists – results, review and recommendations. *Radiology* **139**: 545–9.
22 CARRASCO CH, WALLACE S, RISHLI WR (1991). Percutaneous skeletal biopsy. *Cardiovascular and Interventional Radiology* **14**: 69–72.
23 POPE TL (1987). Letter in *AJR* **149**: 1078.
24 CRAIG FS (1956). Vertebral body biopsy. *Journal of Bone and Joint Surgery* **38A**: 93–102.
25 DAHLBERG PJ (1989). Diagnosis of renal osteodystrophy with the Jamshidi needle biopsy. *Wisconsin Medical Journal* **88**(5): 22–4.
26 MINK JH, WEITZ I, KAGAN AR, STECKEL RJ (1987). Letter/reply. *AJR* **149**: 1078.
27 GIVEN-WILSON R, GISHEN P (1989). Bone biopsy techniques. *Journal of Interventional Radiology* **4**: 59–66.
28 VETRANI A, FULCINITTI F, BOSCHI R, DE ROSA G, PALOMBINI L (1990). Fine needle aspiration biopsy diagnosis of giant cell tumour of bone: An experience with nine cases. *Acta Cytologica* **34**: 863–7.
29 MACLARRON JC (1982). Biopsy of the spine using a needle with a rigid guide wire. *Clinical Radiology* **33**: 189–92.
30 IACCARINO V, SADILE F, VETRANI A, FULCINITTI F, TRONCONE G, RICCIO R, MISASI N (1990). Percutaneous intralesional brushing of cystic lesions of bone: A technical improvement of diagnostic cytology. *Skeletal Radiology* **19**: 198–190.
31 MINK JH (1986). Percutaneous bone biopsy in the patient with known or suspected osseous metastases. *Radiology* **161**: 191–4.
32 TURKEL H, BETHELL FH (1943). Biopsy of bone marrow performed by a new and simple instrument. *Journal of Laboratory and Clinical Medicine* **28**: 1246–51.
33 ACKERMANN W (1956). Vertebral trephine biopsy. *Annals of Surgery* **143**: 373–85.
34 DOMBERNOWSKY P, WORM AH, HAINAN B, HANSEN HH, NISSEN NI (1974). The Radner needle, a bone-marrow biopsy device. *Scandinavian Journal of Haematology* **12**: 270–3.
35 FAUGERE MC, MALLUCHE HH (1983). Comparison of different bone-biopsy techniques for qualitative and quantitative diagnosis of metabolic bone diseases. *Journal of Bone and Joint Surgery* **65A**: 1314–18.
36 MATTHEWS LS, BRAUNSTEIN EM. Corb biopsy needle. Data from Zimmer information brochure.
37 COLLINS JD, BASSETT L, MAIN GD, KAGAN C (1979). Percutaneous biopsy following positive bone scans. *Radiology* **132**: 439–42.
38 MUINDI J, COOMBES RC, GOLDING S, POWLES TJ, KHAN O, HUSBAND J (1983). The role of computed tomography in the detection of bone metastases in breast cancer patients. *British Journal of Radiology* **56**: 233–6.
39 BRUGIERES P, REVEL MP, DUMAS JL, HERAN C, VOISIN MC, GASTON A (1991). CT-guided vertebral biopsy. *Journal of Neuroradiology* **18**: 351–9.
40 COTTY Ph, FOUQUET B, PLESKOF L, ANDURIER A, COTTY F, GOUPILLE Ph, VALAT JP, ALISON D, LAFFONT J (1988).

Vertebral osteomyelitis: Value of percutaneous biopsy. *Journal of Neuroradiology* **15**: 13–21.

41 Thommesen P, Bartholdy N, Fris Anderson MJ (1983). Histiocytosis X: Fine needle aspiration for differentiation between histiocytosis X and spondylitis. *Acta Radiologica. Oncology* **22**: 349–51.

42 Ayala AG, Zornosa J (1983). Primary bone tumours: Percutaneous needle biopsy. *Radiology* **149**: 675–9.

43 Schajowicz F, Derqui JC (1968). Puncture biopsy in lesions of the locomotor system. *Cancer* **21**: 531–48.

44 de Santos L, Murray J, Ayala AG (1979). The value of percutaneous needle biopsy in the management of primary bone tumours. *Cancer* **43**: 735–44.

45 Valls J, Ottolenghi CE, Schajowicz F (1941). La biopsia por aspiración de los cuerpos vertebrales. *Boletines y trabajos, Academia argentina de cirugia* **25**: 743–82.

46 McLaughlin RE, Miller WR, Miller CW (1976). Quadiparesis after needle aspiration of the cervical spine. *Journal of Bone and Joint Surgery* **58A**: 1167–8.

47 Debnam JW, Staple TW (1975). Trephine bone biopsy by radiologists. *Radiology* **116**: 607–9.

48 Ottolenghi CE (1969). Aspiration biopsy of the spine. *Journal of Bone and Joint Surgery* **51A**: 1531–44.

49 Gladstein MO, Grantham SA (1974). Closed skeletal biopsy. *Clinical Orthopaedics and Related Research* **103**: 75–9.

50 Armstrong P, Chalmers AH, Green G, Irvine JD (1978). Needle aspiration/biopsy of the spine in suspected disc space infection. *British Journal of Radiology* **51**: 333–7.

51 de Santos L, Zornoza J (1980). Bone and soft tissue. In *Percutaneous needle biopsy*. Edited by J. Zornoza. Williams & Wilkins, Baltimore.

52 Laredo JD, Bard M (1986). Thoracic spine: Percutaneous trephine biopsy. *Radiology* **160**: 485–9.

53 Odendaal T, Lemmer LB (1991). The value of percutaneous trephine biopsy in the diagnosis of lesions of the vertebral column. *South African Medical Journal* **79**: 21–3.

54 Nagel DA, Albright JA, Keggi KJ, Southwick WO (1965). Closer look at spinal lesions. *Journal of the American Medical Association* **191**(12): 103–6.

4

SPINE

CHAPTER 6

Lumbar zygapophyseal joint pain: diagnostic blocks and therapy

Nikolai Bogduk, Charles Aprill and Richard Derby

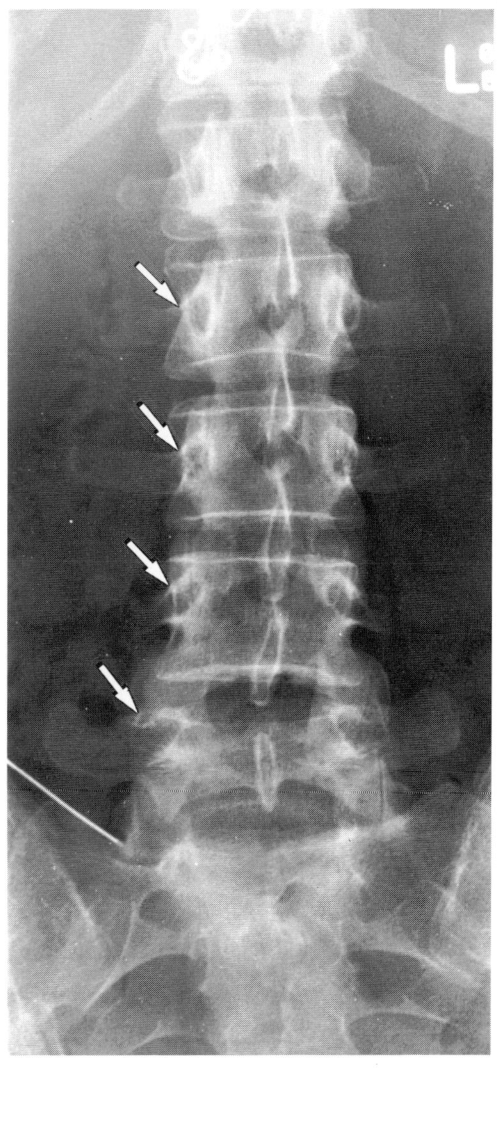

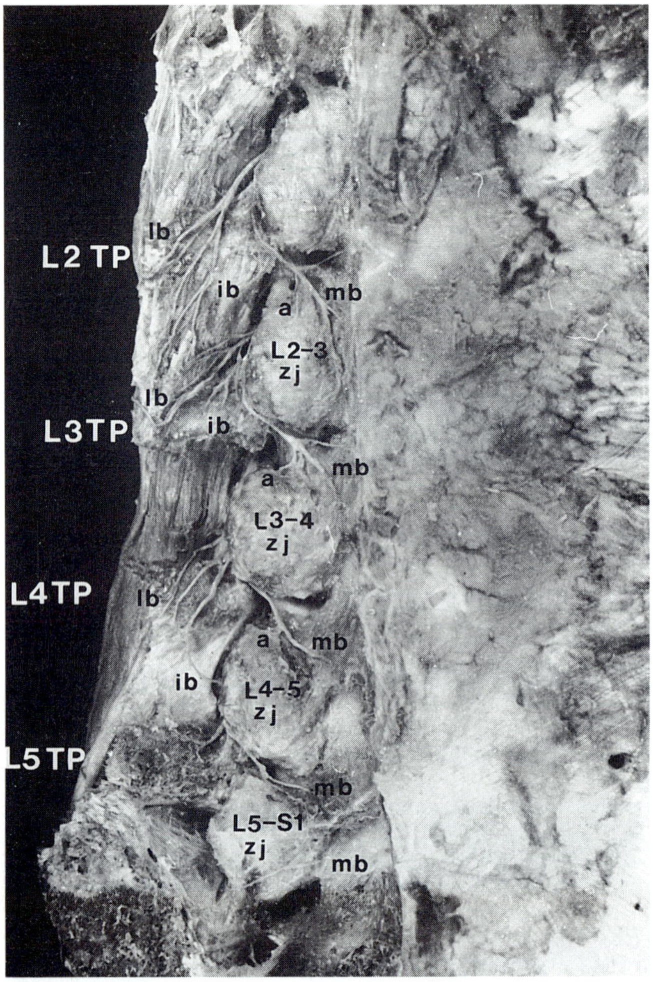

Fig. 6.1
A photograph of a dissection of the branches of the left lumbar dorsal rami. The lateral (lb) and intermediate branches (ib) of the dorsal rami are seen dorsal to the transverse processes (TP). The medial branches (mb) curve medially between the zygapophyseal joints (zj) and furnish articular branches (a) to them.

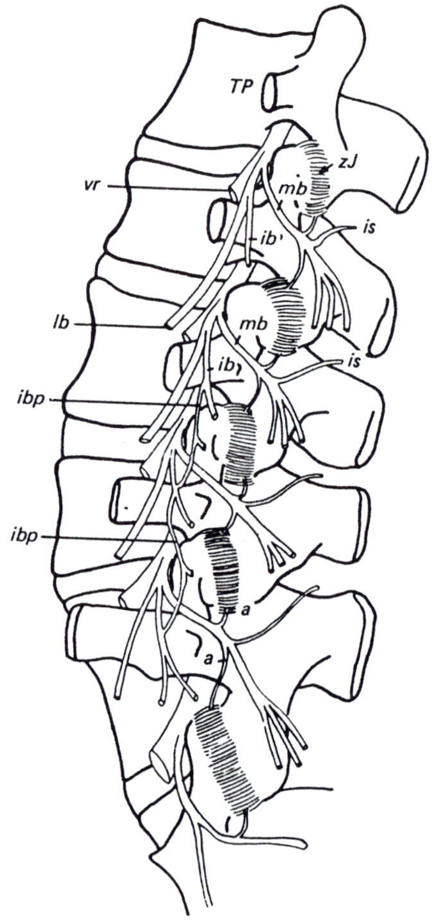

Fig. 6.2
A sketch of a left, postero-lateral view of the branches of the lumbar dorsal rami. The ventral rami (vr) pass in front of the transverse processes (TP). The lateral branches (lb) and intermediate branches (ib) pass behind the transverse processes, the latter forming an intermediate branch plexus (ibp). The medial branches (mb) cross the root of the transverse processes and furnish articular branches to the zygapophyseal joints (zj) before entering the back muscles and supplying the interspinous space (is). Reproduced with permission from Bogduk N, Wilson A S and Tynan W (1982). The human lumbar dorsal rami. *Journal of Anatomy* **134**: 383–97.

The lumbar zygapophyseal joints are paired, synovial joints of the lumbar spine, formed between the inferior articular processes of one vertebra and the superior articular processes of the next, consecutive vertebra. They are innervated by the medial branches of the lumbar dorsal rami.[1] Each medial branch crosses the root of a transverse process and curves medially around the base of a zygapophyseal joint before crossing the lamina below (Figs 6.1 and 6.2). Along its course each nerve furnishes articular branches to the joint above and the joint below, so that each lumbar zygapophyseal joint receives an innervation from the ipsisegmental medial branch and from the segment above (Figs 6.1 and 6.2).

It is abundantly clear that the lumbar zygapophyseal joints are capable of being a source of low back pain and referred pain in the lower limbs. Experiments in normal volunteers have shown that stimulation of these joints with injections of hypertonic saline evokes local pain and pain referred to the ipsilateral buttock and thigh.[2, 3] Several clinical studies have shown that back pain and referred pain in patients can be relieved by anaesthetizing one or more lumbar zygapophyseal joints.[2,4–6]

However, lumbar zygapophyseal joint pain cannot be diagnosed by history or physical examination or radiologically. Although certain clinical features occur more commonly in patients with lumbar zygapophyseal joint pain, no single feature and no combination of features has yet been shown to be pathognomonic of lumbar zygapophyseal joint pain.[4]

Helbig and Lee[5] developed a pre-test scoring system in which points were allocated for the presence or absence of certain putatively diagnostic features. These were back pain associated with groin or thigh pain, localized paraspinal tenderness, reproduction of pain with extension–rotation, and significant corresponding radiographic changes. Patients were then subjected to diagnostic intra-articular lumbar zygapophyseal joint blocks. It transpired that patients scoring over 40 points in their system all responded to diagnostic blocks. However, some responders also scored much less than this. This system, therefore, was highly specific but poorly sensitive; it reliably identifies patients with lumbar zygapophyseal joint pain but fails to detect all such patients.

Some investigators have reported that relief of pain following intra-articular lumbar zygapophyseal joint blocks correlates with the presence of degenerative features evident in CT scans of the joint,[7,8] but the specificity and sensitivity of this relationship is poor; normal-looking joints can also be symptomatic.

The mainstay for diagnosing lumbar zygapophyseal joint pain is diagnostic blocks. In a given patient, a hypothesis can be raised that their pain stems from one or more zygapophyseal joints. Diagnostic blocks can then be used to test this hypothesis. If blocks fail to relieve the patient of their pain, the hypothesis is refuted, but if blocking the joint succeeds in relieving their pain the hypothesis is confirmed.

DIAGNOSTIC BLOCKS

A variety of techniques have been developed whereby the lumbar zygapophyseal joints can be anaesthetized. One approach involves blocking the nerves to the zygapophyseal joints, and has been favoured by neurosurgeons and anaesthetists; the other has focused on the joints themselves and has been favoured by orthopaedic surgeons and radiologists (Fig. 6.3).

The medial branches of the lumbar dorsal rami pursue a constant course in relation to bone.[1] The L1–L4 medial branches regularly cross the junction of the root of the transverse process and the root of the superior articular process; the L5 medial branch crosses the ala of the sacrum adjacent to the root of the S1 superior articular process.[1] These bony sites can be recognized on fluoroscopy, and constitute suitable target points for local anaesthetic blocks of these nerves.[9] By anaesthetizing each of the two medial branches that supply a given lumbar zygapophyseal joint, that joint can be anaesthetized.

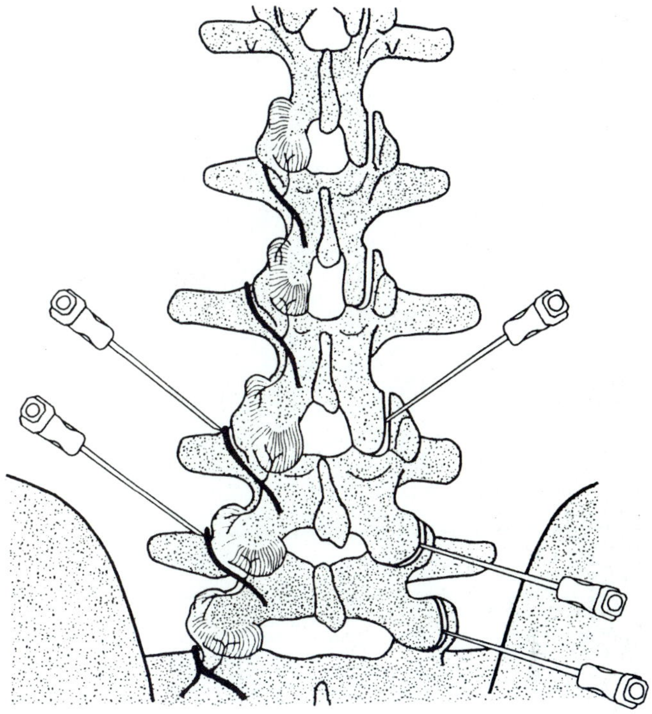

Fig. 6.3
A sketch of a posterior view of the lumbar spine with needles drawn in a position that might be used for the conduct of lumbar medial branch blocks or lumbar intra-articular blocks.

Alternatively, a given lumbar zygapophyseal joint can be anaesthetized directly by infiltrating it, intra-articularly, with local anaesthetic. This approach requires the joint to be entered with a needle and requires arthrography to confirm accurate placement.[4,7,8,10–16]

LUMBAR MEDIAL BRANCH BLOCKS

Lumbar medial branch blocks are performed under aseptic conditions and image intensifier guidance, with the patient lying prone. For the L1–L4 medial branches the target points lie on the dorsal surface of the transverse process immediately below its superior border at its root where it joins the superior articular process.[9,17] In numerical terms, the L4 medial branch crosses the L5 transverse process; the L3 medial branch crosses the L4 transverse process; and so on (Fig. 6.4). The target point for the L5 medial branch lies at the top of the ala of the sacrum at its junction with the S1 superior articular process.[9]

A puncture point on the skin is selected above and lateral to the desired target point, usually above the tip of the target transverse process. This ensures that the needle

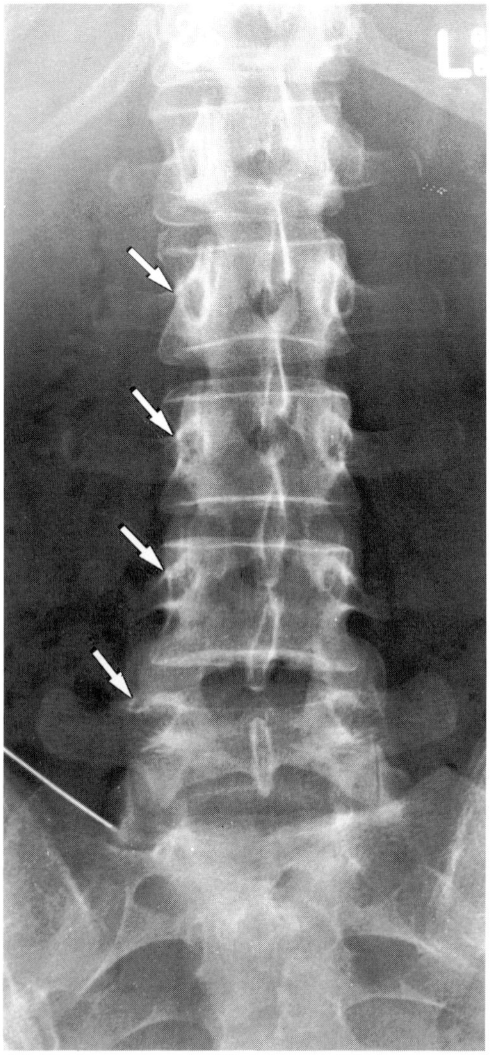

Fig. 6.4
A radiograph of the lumbar spine showing the target points for lumbar medial branch blocks. A needle is seen in position on the ala of the sacrum, in position to block the L5 medial branch.

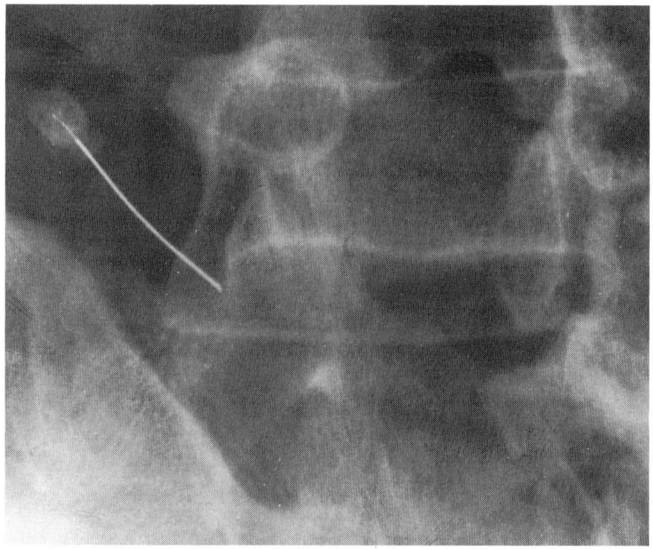

Fig. 6.5
A close-up radiograph of a needle in correct position to anaesthetize the left L4 medial branch where it crosses the transverse process of a transitional L5 vertebra.

will be directed forwards but medially and slightly caudally on to the target point. The medial inclination allows the needle to pass ventral to enlarged mammillary processes which may overhang the target point. The caudad inclination ensures that the injectate is delivered distally on to the target nerve; if the needle is introduced along a rostrad course, the injectate may track towards a spinal nerve and compromise the selectivity of the block.

A 22 G or 25 G 90-mm (3½-in.) spinal needle may be used to perform the block. Larger needles are easier to manoeuvre through the back muscles; finer needles require experience and dexterity to prevent them from diverting from their intended course, but are better tolerated by patients. The needle is introduced through the puncture point and is slowly directed towards the

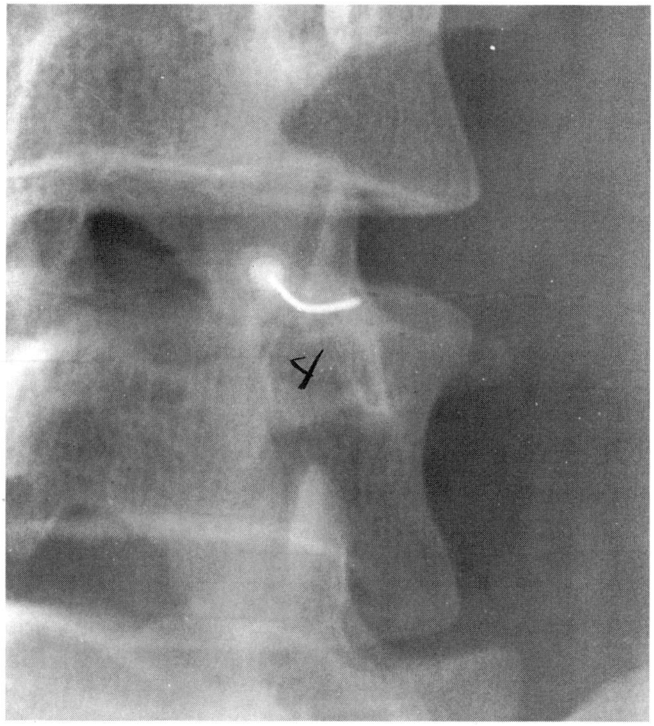

Fig. 6.6
An oblique view of the lumbar spine showing a needle in correct position to block the L3 medial branch where it crosses the L4 transverse process.

target point using periodic screening to maintain course. The needle should be directed first to contact the back of the root of the transverse process; this prevents the needle from being inserted too deeply through the inter-transverse space, and keeps the needle tip clear of the ventral ramus and spinal nerve. Once located on the back of the transverse process, the needle can be readjusted towards the precise target point. Correct location is verified radiographically (Figs 6.4 and 6.5). If required, oblique views may be used to confirm exact placement of the needle on to the junction of the transverse process and the superior articular process (Fig. 6.6).

Once the needle is in correct, final position, local anaesthetic may be injected to infiltrate the nerve. A volume of 0.5 ml is sufficient. Extra volume simply spreads beyond the target point. A precaution is that the injection must be performed slowly, at a rate of about 1 ml per minute). Faster injections or injections under high pressure simply cause the injectate to spread beyond the target point.

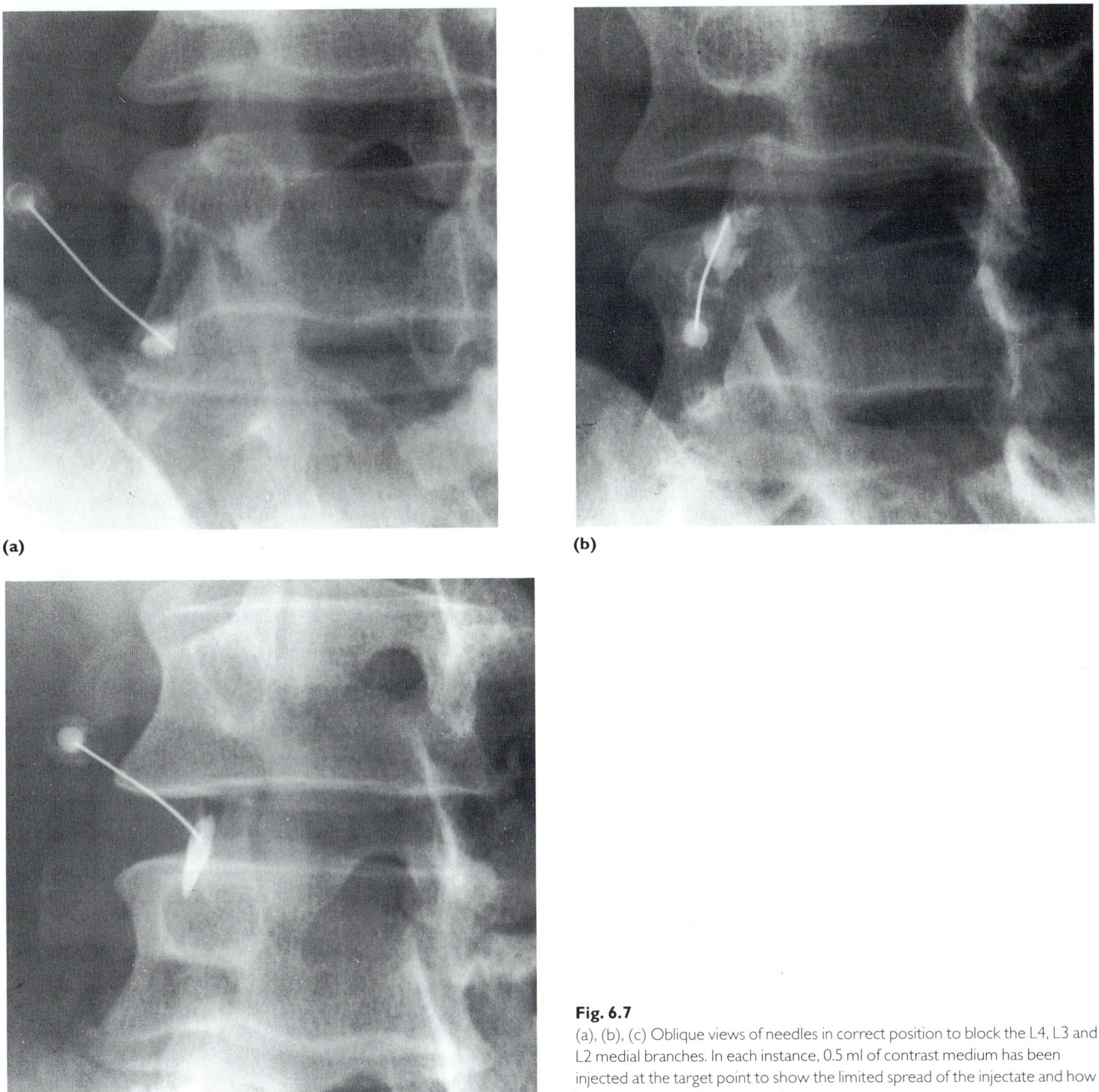

(a) (b) (c)

Fig. 6.7
(a), (b), (c) Oblique views of needles in correct position to block the L4, L3 and L2 medial branches. In each instance, 0.5 ml of contrast medium has been injected at the target point to show the limited spread of the injectate and how it concentrates at the target point.

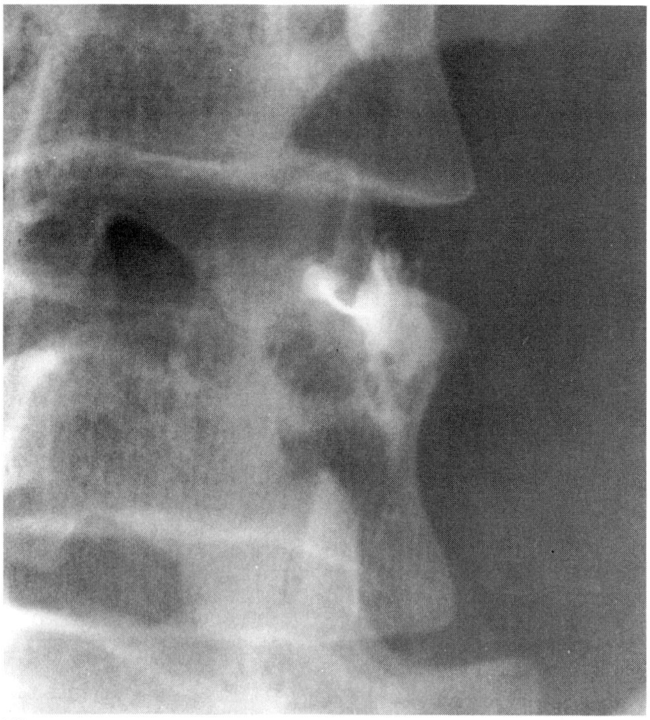

(d)

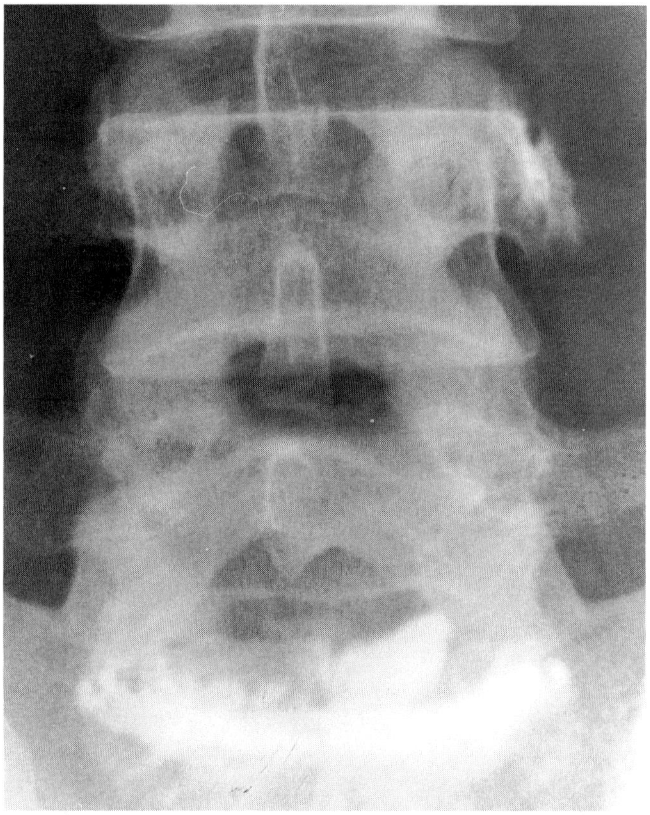

(e)

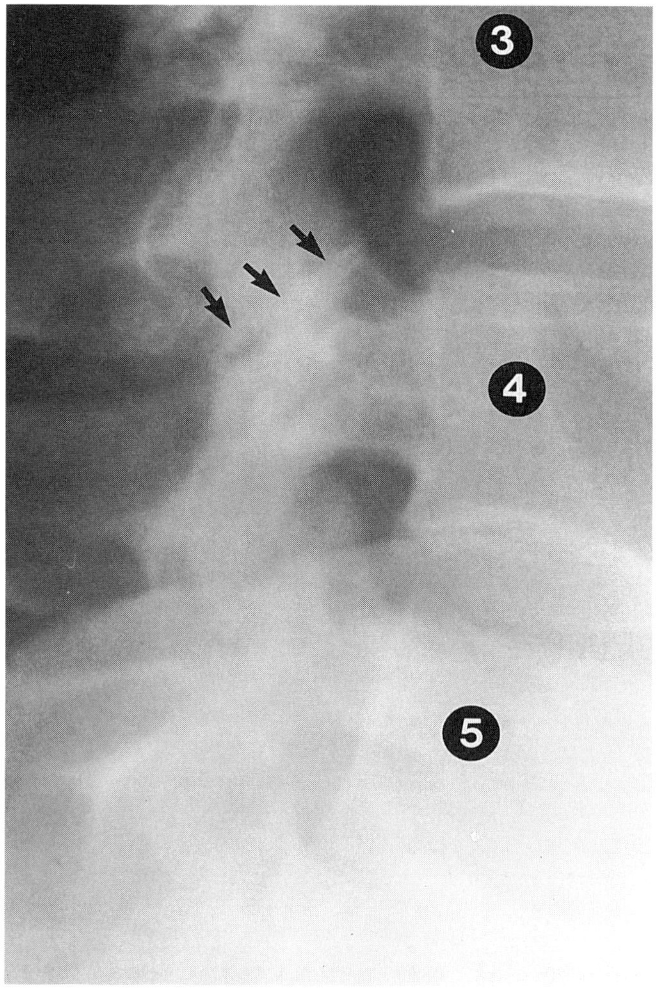

(f)

Fig. 6.7 *Continued*
(d), (e), (f) Oblique, posterior and lateral views of an injection of 0.5 ml of contrast medium on to an L3 medial branch where it crosses the L4 transverse process. In (d) and (e) the contrast medium is evident behind the transverse process along the course of the nerve. In (f) constrast medium is barely evident (arrows), for it forms only a thin film over the neck of the superior articular process but, clearly, it does not extend into the intervertebral foramen or to the spinal nerve.

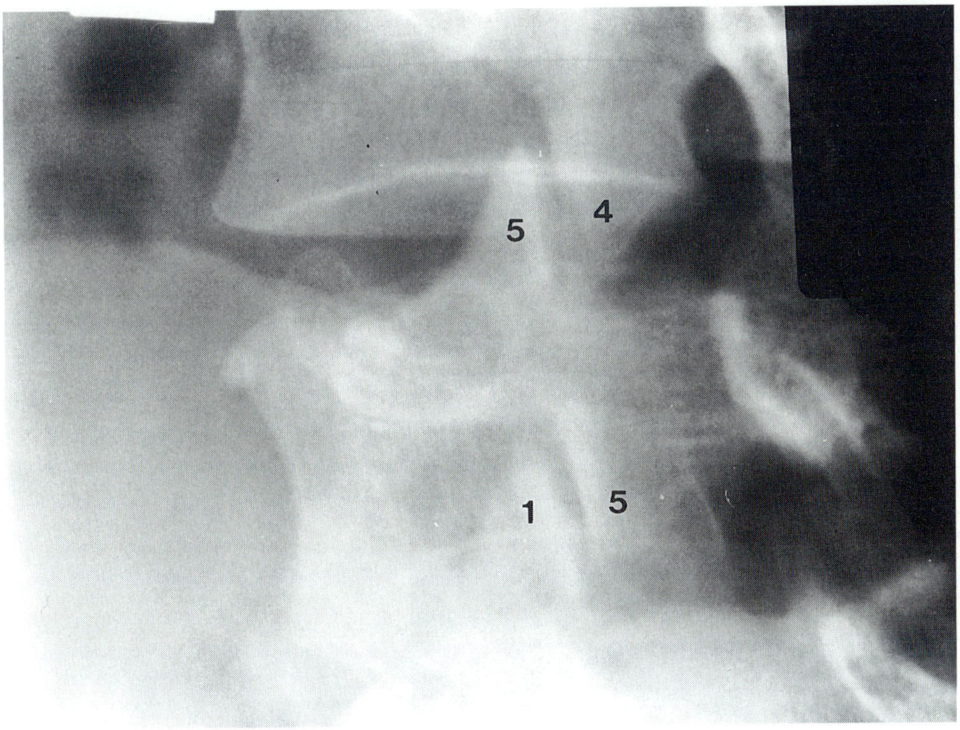

Fig. 6.8
An oblique view of an L4–5 and an L5–S1 zygapophyseal joint, visualizing their cavities.

Injections of contrast medium on to the target point for lumbar medial branch blocks demonstrate that 0.5 ml of injectate fully floods the region where the medial branch crosses the transverse process, and little tracks elsewhere, and certainly not to other structures that might conceivably be sources of pain other than the zygapophyseal joint innervated by the target nerve (Fig. 6.7).

Intra-articular lumbar zygapophyseal joint blocks

Intra articular blocks of the lumbar zygapophyseal joints are performed under image intensifier with the patient initially lying prone. The cavity of the target joint needs to be visualized. Upper lumbar zygapophyseal joints are orientated in the sagittal plane whereupon their cavities are usually quite evident on postero-anterior views of a prone patient. Lower lumbar zygapophyseal joints and particularly the lumbosacral zygapophyseal joints are obliquely orientated, on the average at about 45° to the sagittal plane. To visualize them either the patient has to be rotated appropriately and supported in an oblique, prone position or, if C-arm fluoroscopy is being used, the X-ray beam is tilted to bring the joint space into view (Fig. 6.8).

The skin over the target joint is prepared as for an aseptic procedure, and a puncture point is selected over the joint space as seen on fluoroscopy. A 22 G or 25 G 90-mm (3½-in.) spinal needle is used to gain access to the target joint. Needles of wider gauge offer the advantage that they are more readily directed to the target joint through the overlying back muscles, but difficulties may be encountered in penetrating the joint, particularly if it is narrow. Finer needles enter the joint more easily but are apt to stray during penetration of the back muscles. One variation is to use a double-needle technique in which a 22 G needle is directed on to the joint but a 25 G needle is subsequently passed through it to enter the joint cavity.

The needle is inserted through the puncture point and is directed under periodic image intensifier guidance to the back of the target joint, resting it initially on either the inferior or the superior articular process of the joint to establish the correct depth of insertion. Subsequently, the needle needs only to be readjusted slightly to enter the joint space and needs to be inserted just beyond the initial depth.

The primary target site for entering the joint is at the mid-point of the joint cavity. The needle is aimed at this site and the operator then relies on feel to determine whether the needle enters the joint. The appropriate feel is one of a loss of resistance, changing from that of firm bone to softer tissue (either the joint capsule or its cartilage), with the needle penetrating a few millimetres deeper than the original depth of the articular processes.

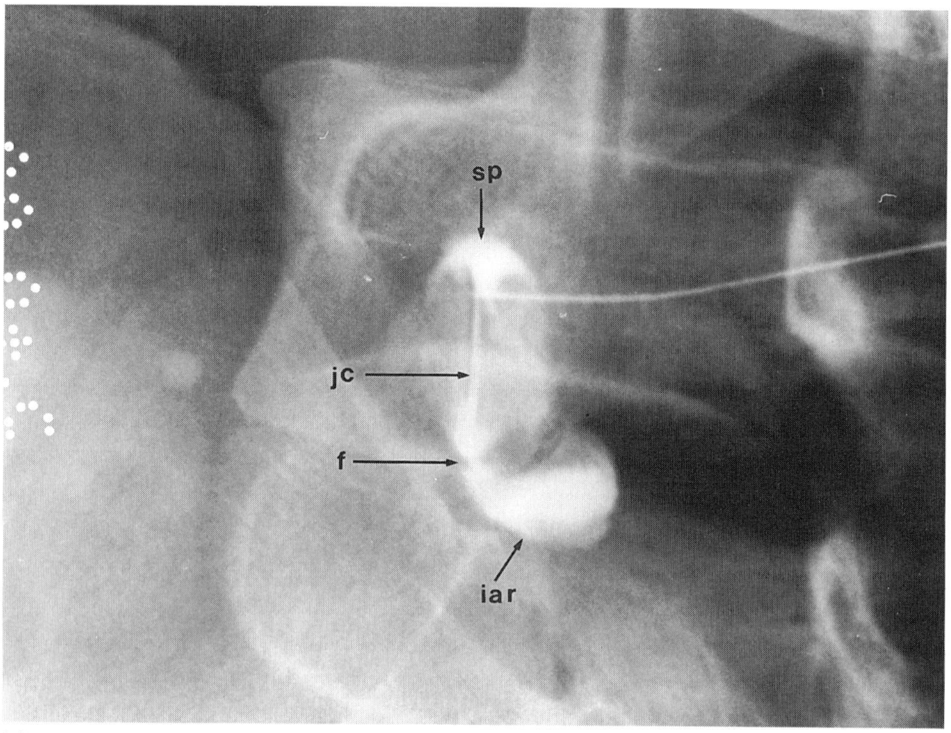

(a)

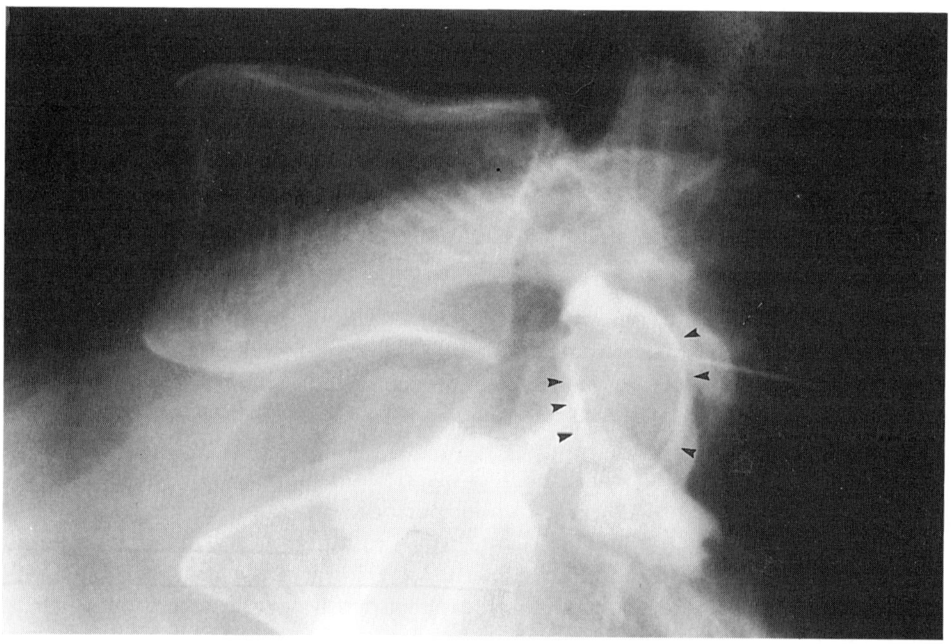

(b)

Fig. 6.9
Oblique (a), lateral (b) and postero-anterior (c) views of an L5–S1 zygapophyseal joint after injection of contrast medium. In (a) the joint cavity (jc) is filled and the contrast medium fills the superior, subcapsular pockets (sp), but inferiorly it escapes through a capsular foramen (f) to fill the extracapsular, inferior articular recess (iar). In (b) and (c) the ventral, dorsal, medium and lateral perimeters of the joint capsule (arrowheads) are outlined by the contrast medium.

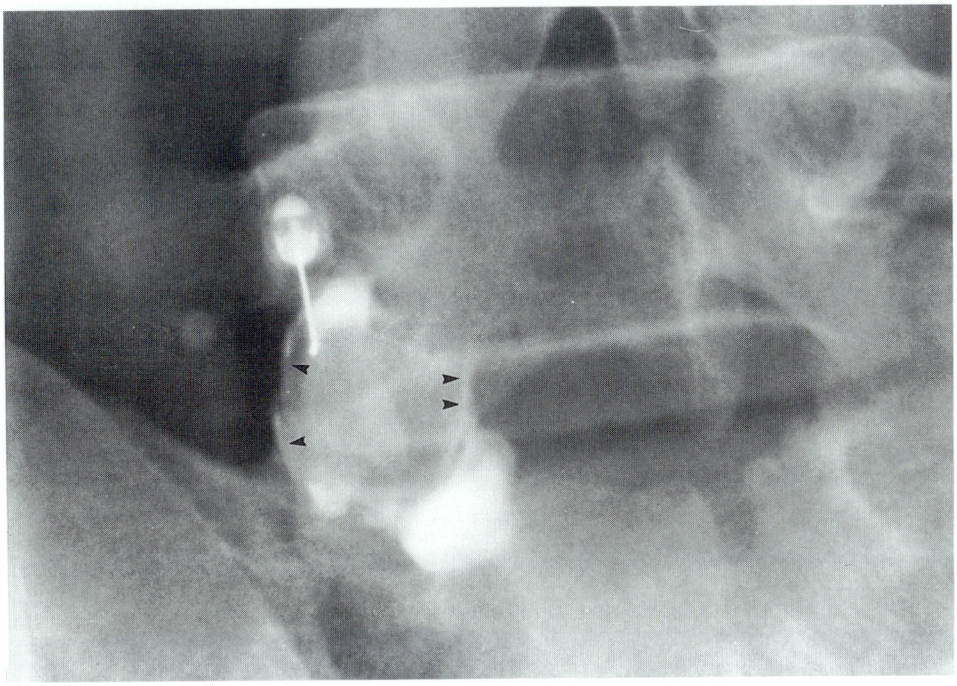

(c)

Fig. 6.9(c)

Authorities differ as to exactly how far the needle should be inserted. Penetrating as far as the centre of the joint space ensures that the joint has been entered, but this penetration may be excessive, and it may become difficult to inject substances if the needle becomes embedded in articular cartilage. It is sufficient otherwise to have the needle simply penetrate the capsule, entering the joint space deep to the capsule but still peripheral to the margins of the articular cartilages.

Once the needle is believed to be inside the joint space, confirmation is achieved by the injection of a minimal quantity (< 0.3 ml) of contrast medium using a small syringe (2–5 ml, but preferably 3 ml syringe) to minimize injection pressure. If the needle is inside the joint, an arthrogram will be achieved, smoothly outlining the perimeter of the joint space (Fig. 6.9).

If the injection has not been made into the joint cavity, contrast medium will be seen to spread beyond the vicinity of the joint in a radiate pattern indicative of extension between the fibres of the multifidus muscle. However, it is quite normal for contrast medium to spread out of the joint cavity through normal capsular foramina into the extracapsular superior and inferior joint recesses (Figs 6.9 and 6.10). Such extension is not a sign of 'extravasation' or joint rupture. Its characteristics are that it lacks a radiate pattern indicative of intramuscular spread, and instead is reminiscent of a lobulated, smooth-edged collection consistent with the fat deposits in the inferior and superior articular recesses having been outlined (Fig. 6.9).

Once intra-articular placement has been verified, local anaesthetic can be injected. To ensure that the joint is not ruptured, not more than 1.0 ml of any agent should be injected.

Selection of procedure

A variety of considerations apply as to whether medial branch blocks or intra-articular blocks should be used as the diagnostic test for lumbar zygapophyseal joint pain. As a screening procedure, medial branch blocks are easier and faster to perform than intra-articular blocks and are equally valid. They avoid the penetration and injection of possibly normal joints when the diagnosis is still uncertain. If required, a positive response to medial branch blocks can later be confirmed by intra-articular blocks.

The choice of procedure may be dictated by the treatment options. Neurosurgeons seeking to treat zygapophyseal joint pain by percutaneous radio frequency neurotomy have no call to perform intra-articular blocks; blocking the nerves to be coagulated is sufficient. On the other hand, if intra-articular therapy or orthopaedic procedures are contemplated, intra-articular blocks become the appropriate diagnostic procedure.

In some situations, gross osteoarthrotic changes may preclude entry into the target joint (Fig. 6.11). The joint may nevertheless be anaesthetized using medial branch blocks.

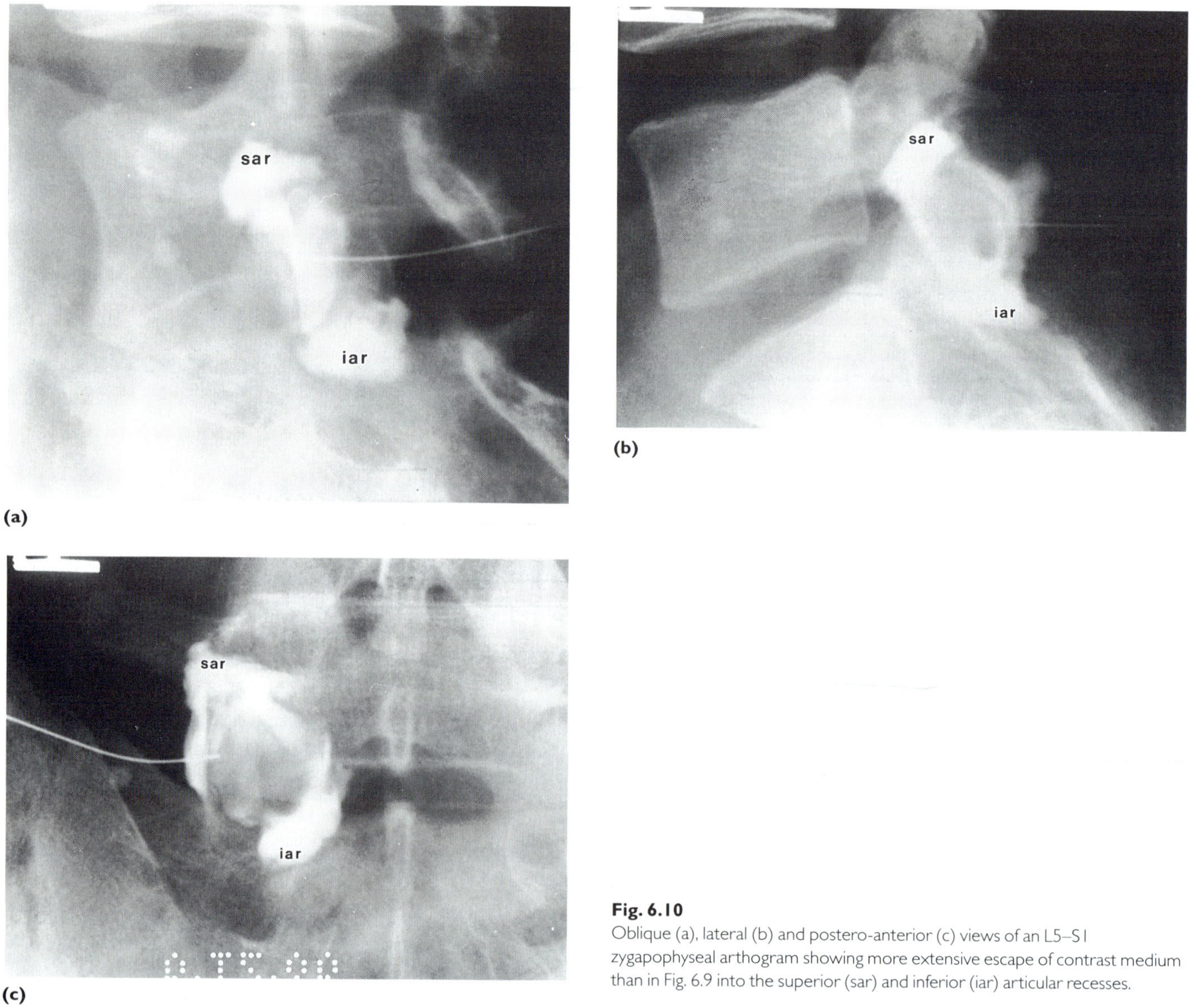

Fig. 6.10
Oblique (a), lateral (b) and postero-anterior (c) views of an L5–S1 zygapophyseal arthogram showing more extensive escape of contrast medium than in Fig. 6.9 into the superior (sar) and inferior (iar) articular recesses.

INTERPRETATION

Like any diagnostic procedure that relies on a patient's subjective response, diagnostic joint blocks are prone to false positive responses. Any diagnostic procedure should therefore be designed to ensure that the response offered by the patient is accurate and in accord with the diagnostic hypothesis being tested.

In the case of zygapophyseal joint blocks, the hypothesis is that the pain stems from the target joint. Consequently, the pain should be relieved completely if the joint is adequately anaesthetized. However, a patient may report a positive response for reasons other than actual relief of pain. The patient may in good faith be wishing for a positive response and suffer a placebo effect. Under these conditions, a single diagnostic block is liable to confounding influences. Thus a firm diagnosis cannot be made on the basis of a single block. A recent study has shown that up to 32 per cent of patients undergoing lumbar zygapophyseal joint blocks report a false-positive response; they obtain the same relief with intramuscular injections of normal saline as with intra-articular injections of local anaesthetic.[18] Consequently, some form of control is required in order to exclude false-positive responses lest a mistaken diagnosis be made and inappropriate therapy instituted.

The most stringent form of control would be injections of normal saline but in most circumstances these might not be ethical. More palatable is a control based on a second, contrasting and confirmatory block. If, on separate occasions, the same joint is anaesthetized on a single-blind or double-blind basis with different local anaesthetic agents, one short-acting (e.g. lignocaine) and

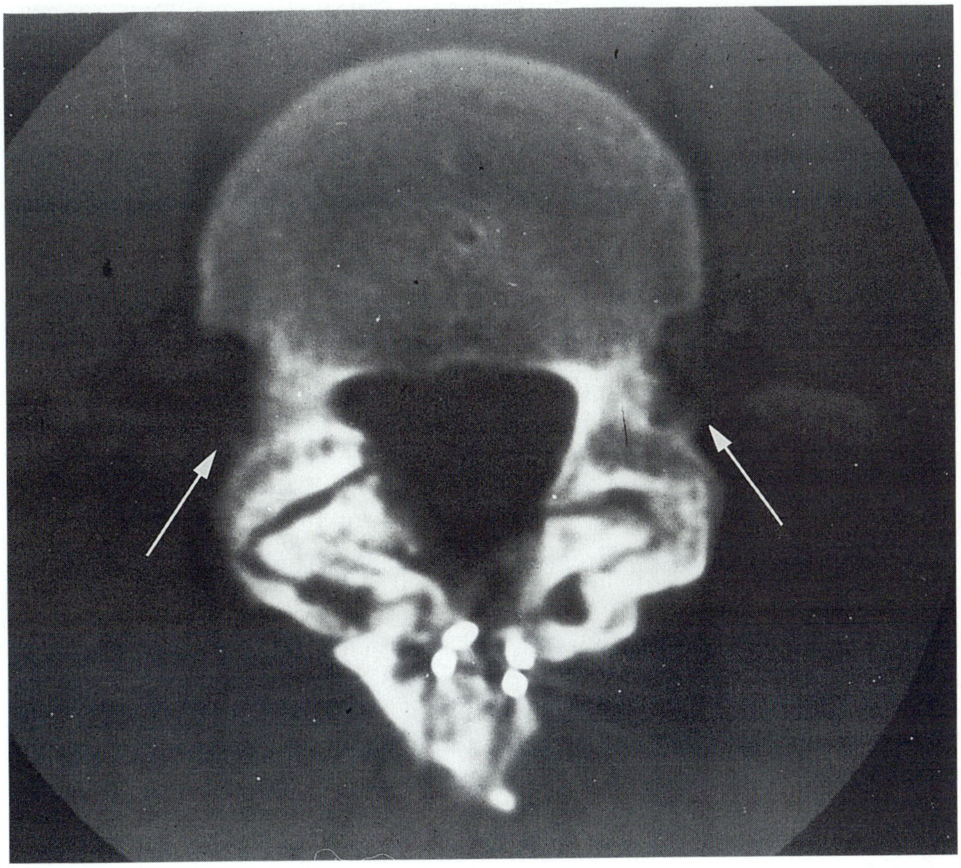

Fig. 6.11
A CT scan showing zygapophyseal joints with severe osteoarthrosis. The cavities of the joints would not be accessible for intra-articular blocks, but pain stemming from them could none the less be blocked by the anaesthetizing of the medial branches that innervate them. The target points for medial branch blocks (arrows) are unobstructed. Courteously provided by Dr Richard North, Department of Neurosurgery, Johns Hopkins Hospital, Baltimore, MD, USA.

one long-acting (e.g. bupivacaine), the genuine patient with a genuine joint complaint should offer differential responses: short-lived relief and long-lasting relief respectively, commensurate with the expected duration of action of the local anaesthetic used. To identify such patients it is mandatory that every patient who offers a positive response to the first block should undergo a second confirmatory block using a different local anaesthetic agent.

Failure to observe the expected pattern of response is not evidence that the patient is a malingerer. The patient may be a poor witness or the patient's pain may not constantly emanate from the target joint. In either case, a firm diagnosis cannot be sustained and should not be entertained. If the response to diagnostic blocks is unreliable, the response to future therapy is likely to be similarly unreliable. Patients with discordant responses should be re-evaluated and not accepted as having an archetypical pain syndrome stemming from the target joint.

Patients may offer a genuine response concordant with the local anaesthetic used, and the relief may be definite but not complete. Such patients warrant further investigation, for the blocks indicate that the injected joint is a contributor to the patient's pain but not its sole source. Such patients may have pain from more than one zygapophyseal joint or from structures in addition to the zygapophyseal joints blocked, such as an intervertebral disc. For complete definition of their source of pain these patients require judicious application of joint blocks performed simultaneously at multiple levels or joint blocks performed in combination with provocation and analgesic discography.

Prevalence

There are no valid data on the prevalence of lumbar zygapophyseal joint pain. Estimates from small and probably contrived samples place the prevalence at around 16–22 per cent,[3,14,16] at around 40 per cent[11,12] or as high as 75 per cent.[8] Larger and less enthusiastic studies placed this figure as low as 7 per cent.[6,19] A valid figure is urgently required, for if one in four low-back patients could be suffering from zygapophyseal joint pain then the use of zygapophyseal joint blocks as a routine procedure is arguably justified; but the same cannot be said if the prevalence is less than 5 per cent. In that event a better screening procedure needs to be developed.

Complications

The only risks of lumbar zygapophyseal joint blocks are those associated with any procedure involving local anaesthetics (and contrast medium): hypersensitivity reactions and infections. The former are rare and have not been reported in the context of lumbar zygapophyseal joint blocks. The latter are avoided by proper attention to aseptic technique.

Treatment

Lumbar zygapophyseal joint pain has been treated in two main ways. The orthopaedic and radiological approach has been to treat the joint pain with intra-articular steroids. The neurosurgical approach has been to denervate the offending joint. Neither approach has been validated.

Intra-articular steroids were introduced as a presumptive therapy for lumbar zygapophyseal joint pain probably on the basis of the alleged beneficial effect of intra-articular steroids for osteoarthritis of the joints of the appendicular skeleton. Proponents of intra-articular steroids for low back pain have reported enthusiastically their successes in open trials, claiming good relief in between 28 per cent and 54 per cent of patients for 3 months or more.[8, 11–13, 15,20] However, reflection on the data reported reveals sobering features.

Patients have been treated with intra-articular steroids frequently without a diagnosis of zygapophyseal joint pain having first been established. The investigator's suspicion or conviction that the patient had zygapophyseal joint pain was the only indication to proceed with therapy. Consequently, in highly selected patients the results appear good; in more randomly selected patients the results are less impressive.

The one carefully designed, double-blind study of intra-articular steroids for lumbar zygapophyseal joint pain has produced striking results.[6] Before entering the trial all patients were screened with intra-articular injections of local anaesthetic. Only responders entered the trial. Subsequently, comparison of 48 patients treated with intra-articular steroids and 48 patients treated with intra-articular normal saline revealed no clinically significant differences at follow-up 1 month and 6 months later.

This study did not deny the existence of lumbar zygapophyseal joint pain. It showed only that intra-articular steroids offer no particular benefit over intra-articular normal saline. Notwithstanding this result, it is notable that in the study an appreciable number of patients reported benefit regardless of the agent used. This may indicate that some other form of intra-articular therapy may prove to be beneficial for lumbar zygapophyseal joint pain, but, at this stage, intra-articular steroids do not appear to be the definitive answer.

A confounding aspect of this study, however, is that very few of the patients treated had obtained complete relief of their back pain following diagnostic blocks. Thus, the majority of patients who entered the trial may not have suffered 'pure' zygapophyseal joint pain; they may have had other sources of pain that prevented them obtaining relief following intra-articular steroids. The study, therefore, does not necessarily reflect the true efficacy of intra-articular steroids in patients with isolated zygapophyseal joint pain.

Some investigators have explored the use of phenol injected on to the medial branches of the lumbar dorsal rami.[21,22] The early use of this therapy was attended by modest results;[21] a more enthusiastic report has appeared more recently.[22] However, although long follow-up periods were reported in this latter study, no controls were included.

Surgeons and anaesthetists have otherwise explored the use of percutaneous radio frequency denervation in the treatment of zygapophyseal joint pain. However, what has polluted and confounded this field and given it a bad name are the poor techniques that have been used. The initial enthusiastic reports of Shealy[23–26] became meaningless once it was demonstrated that the target points he advocated and used for lumbar facet denervation were nowhere near the appropriate nerves[9, 27] Similarly, all those studies which used the Shealy approach are invalid, save only to serve as a warning that good results can be reported even if essentially sham lesions are used in the treatment of back pain[28–38]

Yet even those studies which subsequently used anatomically correct target points must now face scrutiny.[39–41] Technical studies have demonstrated that radio frequency electrodes do not coagulate beyond their tip; they coagulate circumferentially.[42] Consequently, electrodes introduced orthogonally on to a nerve (the way diagnostic blocks are performed) are very likely to fail to coagulate the nerve adequately. Short-term benefits only can be expected, if any at all, which is consistent with the reasonable but limited good results reported at early follow-up and the decay of responses by 1 year.[39–41]

To achieve thorough coagulation, the electrode must be introduced across or parallel to the target nerve, but to date this correct approach has not been evaluated in clinical trials, either open or controlled.

Percutaneous radio frequency neurotomy remains a plausible means of treatment for lumbar zygapophyseal joint pain. The poor results of previous studies can be explained on the basis of technical failure. Consequently, the opportunity exists for resetting this field on a proper scientific foundation. Radio frequency neurotomy lends itself superbly to double-blind trials. The operator can introduce the electrode with care and conviction but can remain masked as to whether active or inactive current is passed by an assistant who controls the radio frequency generator. Such trials should be performed for lumbar medial branch neurotomy, lest another potentially sham procedure be wantonly perpetuated amongst irresponsible and uninformed clinicians in this field.

References

1 Bogduk N, Wilson AS, Tynan W (1982). The human lumbar dorsal rami. *Journal of Anatomy* **134**: 383–97.

2 Mooney V, Robertson J (1976) The facet syndrome. *Clinical Orthopaedics and Related Research* **115**: 149–56.

3 McCall IW, Park WM, O'Brien JP (1979). Induced pain referral from posterior lumbar elements in normal subjects. *Spine* **4**: 441–6.

4 Fairbank JCT, Park WM, McCall IW, O'Brien JP (1981). Apophyseal injection of local anaesthetic as a diagnostic aid in primary low-back pain syndromes. *Spine* **6**: 598–605.

5 Helbig T, Lee CK (1988). The lumbar facet syndrome. *Spine* **13**: 61–4.

6 Carette S, Marcoux S, Truchon R, Grondin C, Gagnon J, Allard Y, Latulippe M (1991). A controlled trial of corticosteroid injections into facet joints for chronic low back pain. *New England Journal of Medicine* **325**: 1002–7.

7 Carrera GF, Williams AL (1984). Current concepts in evaluation of the lumbar facet joints. CRC *Critical Reviews in Diagnostic Imaging* **21**: 85–104.

8 Lewinnek GE, Warfield CA (1986). Facet joint degeneration as a cause of low back pain. *Clinical Orthopaedics and Related Research* **213**: 216–22.

9 Bogduk N, Long DM (1980). Percutaneous lumbar medial branch neurotomy. A modification of facet denervation. *Spine* **5**, 193–200.

10 Dory MA (1981). Arthrography of the lumbar facet joints. *Radiology* **140**: 23–7.

11 Lau LSW, Littlejohn GO, Miller MH (1982). Clinical evaluation of intra-articular injections for lumbar facet joint pain. *Medical Journal of Australia* **143**: 563–5.

12 Lippit AB (1984). The facet joint and its role in spine pain: management with facet joint injections. *Spine* **9**: 747–50.

13 Lynch MC, Taylor JF (1986). Facet joint injection for low back pain. *Journal of Bone and Joint Surgery* **68B**: 138–41.

14 Moran R, O'Connell D, Walsh MG (1988). The diagnostic value of facet joint injections. *Spine* **12**: 1407–10.

15 Murtagh FR (1988). Computed tomography and fluoroscopy guided anaesthesia and steroid injection in facet syndrome. *Spine* **13**: 686–9.

16 Raymond J, Dumas JM (1984). Intra-articular facet block: Diagnostic test or therapeutic procedure? *Radiology* **151**: 333–6.

17 Bodguk N (1988). Back pain: Zygapophysial blocks and epidiural steroids. In: *Neural Blockade in Clinical Anesthesia and Management of Pain*, pp. 935–54. Edited by Cousins MJ and Bridenbaugh PO. Lippincott, Philapdelphia.

18 Schwartzer AC, Wang S, Laurent R, McNaught P, Brooks PM (1992). The role of the zygapophysial joint in chronic low back pain. *Australian and New Zealand Journal of Medicine* **22**: 185.

19 Jackson RP, Jacobs RR, Montesano PX (1988). Facet joint injection in low-back pain: A prospective statistical study. *Spine* **13**: 966–71.

20 Destouet JM, Gilula LA, Murphy WA, Monsees B (1982). Lumbar facet joint injection: Indication, technique, clinical correlation, and preliminary results. *Radiology* **145**: 321–5.

21 Hickey RFJ, Tregonning GD (1977). Denervation of spinal facets for treatment of chronic low back pain. *New Zealand Medical Journal* **85**: 96–9.

22 Silvers HR (1990). Lumbar percutaneous facet rhizotomy. *Spine* **15**: 36–40.

23 Shealy CN (1974a). Facets in back and sciatic pain. *Minnesota Medicine* **57**: 199–203.

24 Shealy CN (1974b). The role of the spinal facets in back and sciatic pain. *Headache* **14**: 101–4.

25 Shealy CN (1975). Percutaneous radiofrequency denervation of spinal facets. *Journal of Neurosurgery* **43**: 448–51.

26 Shealy CN (1976). Facet denervation in the management of back sciatic pain. *Clinical Orthopaedics and Related Research* **115**: 157–64.

27 Bogduk N, Long DM (1979). The anatomy of the so-called 'articular nerves' and their relationship to facet denervation in the treatment of low back pain. *Journal of Neurosurgery* **51**: 172–7.

28 Andersen KH, Mosdal C, Vaernet K (1987). Percutaneous radiofrequency facet denervation in low-back and extremity pain. *Acta Neurochirugica* **87**: 48–51.

29 Banerjee T, Pittman HH (1976). Facet rhizotomy: Another armamentarium for treatment of low backache. *North Carolina Medical Journal* **37**: 354–60.

30 Fassio B, Bouvier JP, Ginestie JF (1981). Dénervation articulaire postérieure per-cutanée et chirurgicale. *Revue de la Chirurgie Orthopédique* Supplement II, 131–6.

31 Florez G, Erias J, Ucar S (1977). Percutaneous rhizotomy of the articular nerve of Luschka for low back and sciatic pain. *Acta Neurochirurgica Supplementum* **24**: 67–71.

32 Fuentes E (1978). La neurotomía apofisaria transcutánea en el tratamiento de la lumbalgia crónica. *Revista Médica de Chile* **106**: 440–3.

33 Lora J, Long DM (1976). So-called facet denervation in the management of intractable back pain. *Spine* **1**: 121–6.

34 McCulloch JA, Organ LW (1977). Percutaneous radiofrequency lumbar rhizolysis (rhizotomy). *Canadian Medical Association Journal* **116**: 30–2.

35 Oudenhoven RC (1974). Articular rhizotomy. *Surgical Neurology* **2**: 275–8.

36 Pawl RP (1974). Results in the treatment of low back syndrome from sensory neurolysis of lumbar facets (facet rhizotomy) by thermal coagulation. *Proceedings of the Institute of Medicine of Chicago* **30**: 150–1.

37 Schaerer JP (1978). Radiofrequency facet rhizotomy in the treatment of chronic neck and low back pain. *International Surgery* **63**: 53–9.

38 Uyttendaele D, Verhamme J, Vercauteren M (1981). Local block of lumbar facet joints and percutaneous

radiofrequency denervation. Preliminary results. *Acta Orthopaedica Belgica* **47**: 135–9.

39 MEHTA M, SLUIJTER ME (1979). The treatment of chronic low back pain. *Anaesthesia* **34**: 768–75.

40 OGSBURY JS, SIMON RH, LEHMAN RAW (1977). Facet denervation in the treatment of low back syndrome. *Pain* **3**: 257–63.

41 RASHBAUM RF (1983). Radiofrequency facet denervation. A treatment alternative in refractory low back pain with or without leg pain. *Orthopedic Clinics of North America* **14**: 569–75.

42 BOGDUK N, MACINTOSH J, MARSLAND A (1987). Technical limitations to the efficacy of radiofrequency neurotomy for spinal pain. *Neurosurgery* **20**: 529–35.

ADDENDUM

In the time since this chapter was originally composed several studies have been completed that address the prevalence of zygapophyseal joint pain, its clinical and radiological features, and its relationship to discogenic pain. The reader is referred to these publications for the latest information on these issues.

1 SCHWARZER AC, APRILL CN, DERBY R, FORTIN J, KINE G, BOGDUK N (1994). Clinical features of patients with pain stemming from the lumbar zygapophysial joints. Is the lumbar facet syndrome a clinical entity? *Spine* **19**:1132–7.

2 SCHWARZER AC, APRILL CN, DERBY R, FORTIN J, KINE G, BOGDUK N (1994). The relative contributions of the disc and zygapophyseal joint in chronic lower back pain. *Spine* **19**:801–6.

3 SCHWARZER AC, APRILL CN, DERBY R, FORTIN J, KINE G, BOGDUK N (1994). The false-positive rate of uncontrolled diagnostic blocks of the lumbar zygapophysial joints. *Pain* **58**:195–200.

4 SCHWARZER AC, DERBY R, APRILL CN, FORTIN J, KINE G, BOGDUK N (1994). Pain from the lumbar zygapophysial joints: a test of two models. *J Spinal Disord* **7**:331–6.

5 SCHWARZER AC, DERBY R, APRILL CN, FORTIN J, KINE G, BOGDUK N (1994). The value of the provocation response in lumbar zygapophyseal joint injections. *Clinical Journal of Pain* **10**: 309–13.

6 SCHWARZER AC, WANG S, BOGDUK N, MCNAUGHT PJ, LAURENT R (1995). The prevalence and clinical features of lumbar zygapophysial joint pain: a study in an Australian population with chronic low back pain. *Annals of the Rheumatic Diseases* (in press).

7 SCHWARZER AC, WANG S, O'DRISCOLL D, HARRINGTON T, BOGDUK N, LAURENT R (1995). The ability of computed tomography to identify a painful zygapophysial joint in patients with chronic low back pain. *Spline* (in press).

CHAPTER 7

Discography

Iain McCall

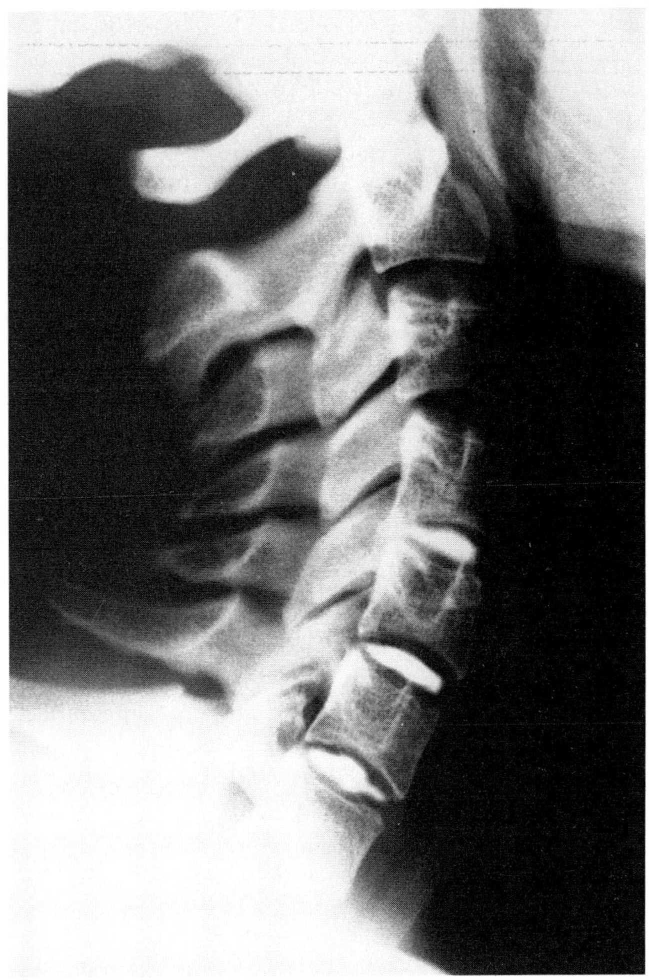

The examination of the internal structure of the intervertebral disc by means of the insertion of contrast medium into the nucleus pulposus was first described by Lindblom.[1] Since his initial studies the technique has evolved and, despite many advances in imaging methods, it remains a valuable adjunct to the assessment of patients with neck or low back pain, with or without associated limb pain.

LUMBAR DISCOGRAPHY

Indications

Discography is primarily a means whereby the exact level of symptomatic pain may be identified in a patient with long-standing back and/or referred leg pain who has not responded to comprehensive conservative therapy. It is usually performed as a prelude to a fusion of the relevant spinal segment. It may be used to assess the status of other discs, so that the extent of the fusion may be clearly defined. Discography has also been used in some series[2] to confirm a disc prolapse causing sciatica, but this is not a routine use of the technique, as the non-invasive techniques of CT or MRI are the initial investigation of choice for this condition.

Technique

PREPARATION AND INFORMED CONSENT

The patient will require a detailed explanation of the procedure and must be clearly informed that it is uncomfortable. The patient should be told that pain may occur and that they should inform the radiologist where the pain is being felt. The patient should not be told prior to the procedure that the symptoms may be reproduced as this will reduce the objectivity of the procedure and the patient will feel that the test has been a failure if the pain has not been reproduced. The radiologist should ascertain the pain distribution of the patient prior to the procedure. The patient should also be warned about the risk of inducing a discitis and be questioned for history of allergy or sepsis, either local or distant.

One hour prior to the investigation the patient may be given 10 mg of diazepam orally. A single intravenous injection of antibiotic may also be administered just prior to the procedure, although an alternative system is to inject antibiotics into the disc at the time of the injection of contrast agent.[3] A broad-spectrum cephalosporin has been suggested as appropriate.

The patient may lie on their side or be positioned prone on the screening table. A pad should be placed under the flank to straighten the spine and a mild degree of flexion will assist access to the L5/S1 level. Biplanar screening is necessary and, in the absence of a biplanar screening table, this may be achieved simply with a C-arm portable system rotating between the two planes. Alternatively, a C-arm may be placed at the head of a normal screening table at right angles to the plane of screening of the table and the patient driven upward on the table top to check the needle position.

All effort must be made to reduce the risk of infection. Careful skin preparation is essential and attention to all details of aseptic procedure must be undertaken. Whether the patient is on their side or prone, the approach should be postero-lateral. The needle should be angled between 45 and 60° to the sagittal plane. This is usually achieved by inserting the needle one hand's breadth away from the spinous processes, but some adaptation may be necessary in very slim or obese patients (Fig. 7.1a). The needle should be inserted at the level of the disc to be investigated. This, however, is rarely possible for L5/S1. At this latter level the iliac crest should be palpated and the needle inserted just above the crest at the point of a hand's breadth out from the mid-line. The needle will require angling downwards as well as inward (Fig. 7.1b).

A double-needle approach is utilized by most practitioners of discography, the outer needle being of 21 G and 5 in. (130 mm) in length, the inner being 26 G and 6¼ in. (160 mm) in length. Local anaesthetic is inserted into the skin and into the superficial muscle fascia. The 21 G needle is then inserted and advanced down to the postero-lateral outer annulus. The direction of the needle is checked by screening in both planes and care must be taken to avoid the emerging nerve root. This is more likely to be avoided by directing the needle towards the inferior part of the disc.

The 21 G needle will feel the resistance of the outer annulus and the stilette should then be removed. The 26 G needle is then advanced into the centre of the nucleus pulposus. The 21 G needle should be in the posterior third of the disc on the lateral view. If it is positioned too far anteriorly on the disc, the 26 G needle will pass along the margin of the nucleus into the anterior annulus. This is particularly the case at L5/S1, where the coronal orientation of the facets makes the approach more difficult. In these circumstances the 26 G needle may be curved and inserted with the concavity of the curve facing posteriorly, which may bring the needle tip into the centre of the nucleus. The final position of the needle should be checked on the AP and lateral screening before injection of contrast medium.

An alternative technique is to position the needle under CT control. The overall technique remains the same except that the CT exposures are used to monitor the 21 G needle placement. This technique enables the nerve root to be clearly identified and avoided. It is, however, more time consuming as the patient usually has to be moved in and out of the gantry for needle adjustment.

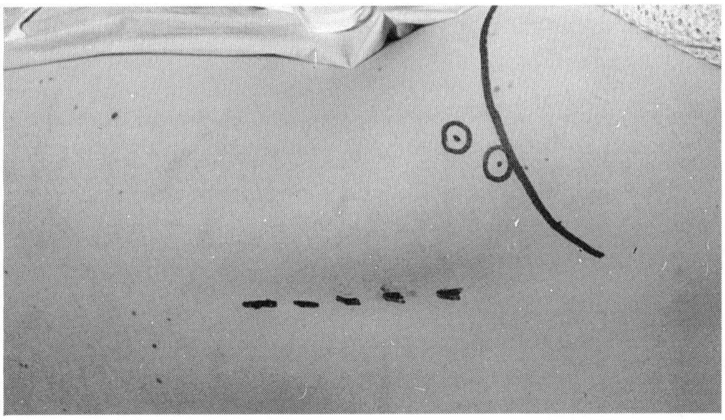

(a)

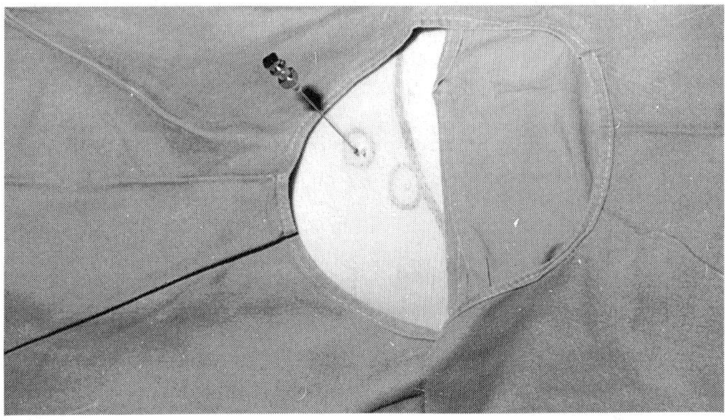

(b)

Fig. 7.1
(a) The site of entry for L4/5 and L5/S1. Disc puncture is demonstrated. The former is one hand's breadth lateral to the spinous process (dotted line). (b) The patient is carefully prepared and covered in sterile drapes. The 21 G needle is angled at 45° to the horizontal.

Finally, the L5/S1 disc may not be accessible from the postero-lateral approach and a transdural mid-line placement may be undertaken. The technique is the same as for a lumbar puncture, with the 21 G needle advanced to the posterior annular margin and then the 26 G needle inserted into the nucleus pulposus.

Injection

Once the needle is in position, the injection of contrast agent may take place. It is important not to use a sodium-based contrast medium as the sodium is neurotoxic and will induce pain even if no pain was originally present. Most centres now use non-ionic water-soluble contrast agents, and this is recommended, as discodural fistulas have been recorded. Some authors have advocated the use of pressure-measuring systems at the time of injection to assist in depicting the pattern of discogenic damage. These, however, are not routinely used in most centres but may be valuable for research purposes.[4] A similar effect may be achieved by injecting the maximum

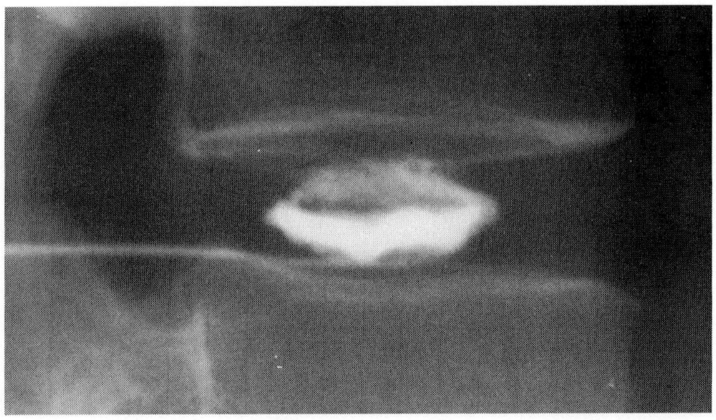

(a)

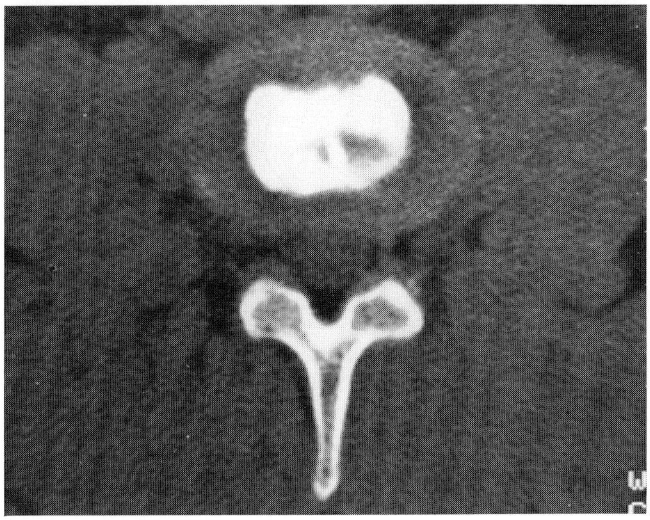

(b)

Fig. 7.2 *Normal discogram*
(a) The contrast fills the centre of the nucleus and contains a normal central cleft. (b) CT shows the intact annulus around the contrast outlined nucleus.

amount of contrast agent that the disc will hold, which will vary with the state of the disc. The injected disc is radiographed in the AP plane and in flexion and extension lateral planes. The exact extent of disc damage can be demonstrated by undertaking CT examinations of the disc following the injection. On the basis of standard AP and lateral views it appears that the normal nucleus will usually accept only a limited amount of contrast with a maximum volume of 2 ml. The injection is made using a 1-ml syringe in order to achieve higher pressure with control. The normal pattern of the injection in a young person is a well-defined collection of contrast in the centre of the nucleus, which is described as a 'cotton ball effect'. This is caused by contrast displacing nuclear material. An alternative normal appearance is the appearance of two convex horizontal curves of contrast with a void in the centre of the so-called hamburger pattern (Fig. 7.2a), with the margins contained within the margins of the nucleus pulposus.[5] The resistance to

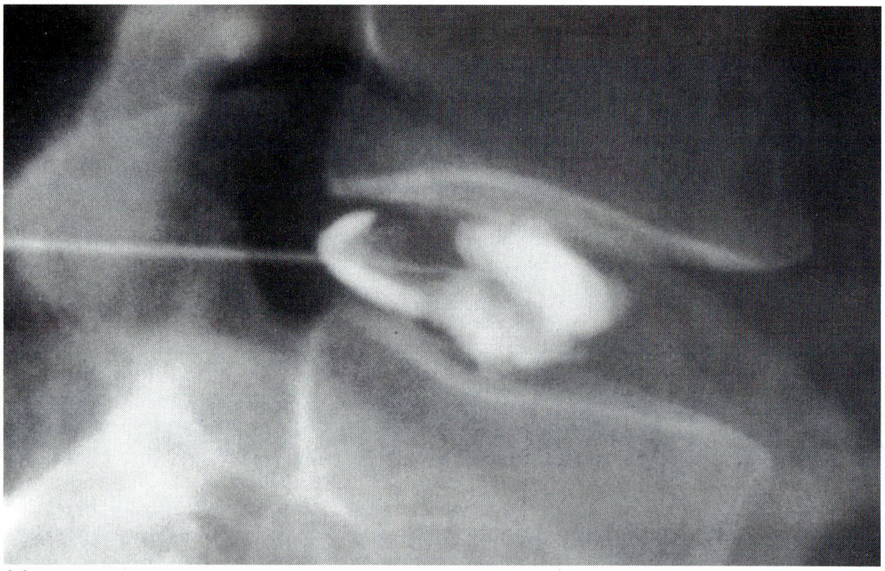

(a)

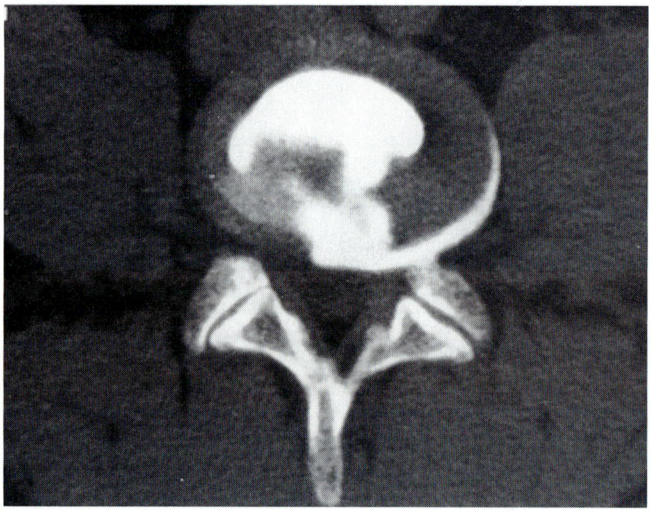

(b)

Fig. 7.3

(a) Internal annular disruption is demonstrated with contrast extending to the outer annular margin. (b) The CT shows the internal derangement of the posterior annulus but also demonstrates the concentric cleft around the periphery of the annulus on the left, giving a grading of 2 on each scoring system.

the injection will gradually increase during the course of the injection. On the CT scan, the contrast is demonstrated as a collection of high attenuation in the centre of the disc (Fig. 7.2b).

Abnormal appearances are well demonstrated on the AP and lateral planes, and additional changes can also be appreciated on the flexion and extension lateral views. The changes on CT can also be graded by the separate assessment of the degeneration that identifies the circumferential extent of disc disruption and annular disruption based on the depth of penetration of the annulus after contrast injection. Degeneration is rated 0 when contrast is restricted to the nucleus pulposus and from 1 to 3 depending on the percentage of annulus filled with contrast, 1 being less than 10 per cent, 2 partial (less than 50 per cent) and 3 total (greater than 50 per cent). Annular disruption is again graded 0 when contrast limited to the nucleus is nil, whereas less than 50 per cent penetration of annular thickness is grade 1, more than 50 per cent but still not complete is graded 2, and full thickness is graded 3.[6]

Patterns of disc damage

1 Minor damage is confined to internal annular disruption, which is characterized on the plain radiographs by lines of contrast seen within the annulus but not extending beyond the confines of the annular margins. These may occur anywhere around the annulus but are more common posteriorly.[7] They would be lower grades of both the degeneration and annular penetration of the discs on CT scans (Fig. 7.3 a and b).

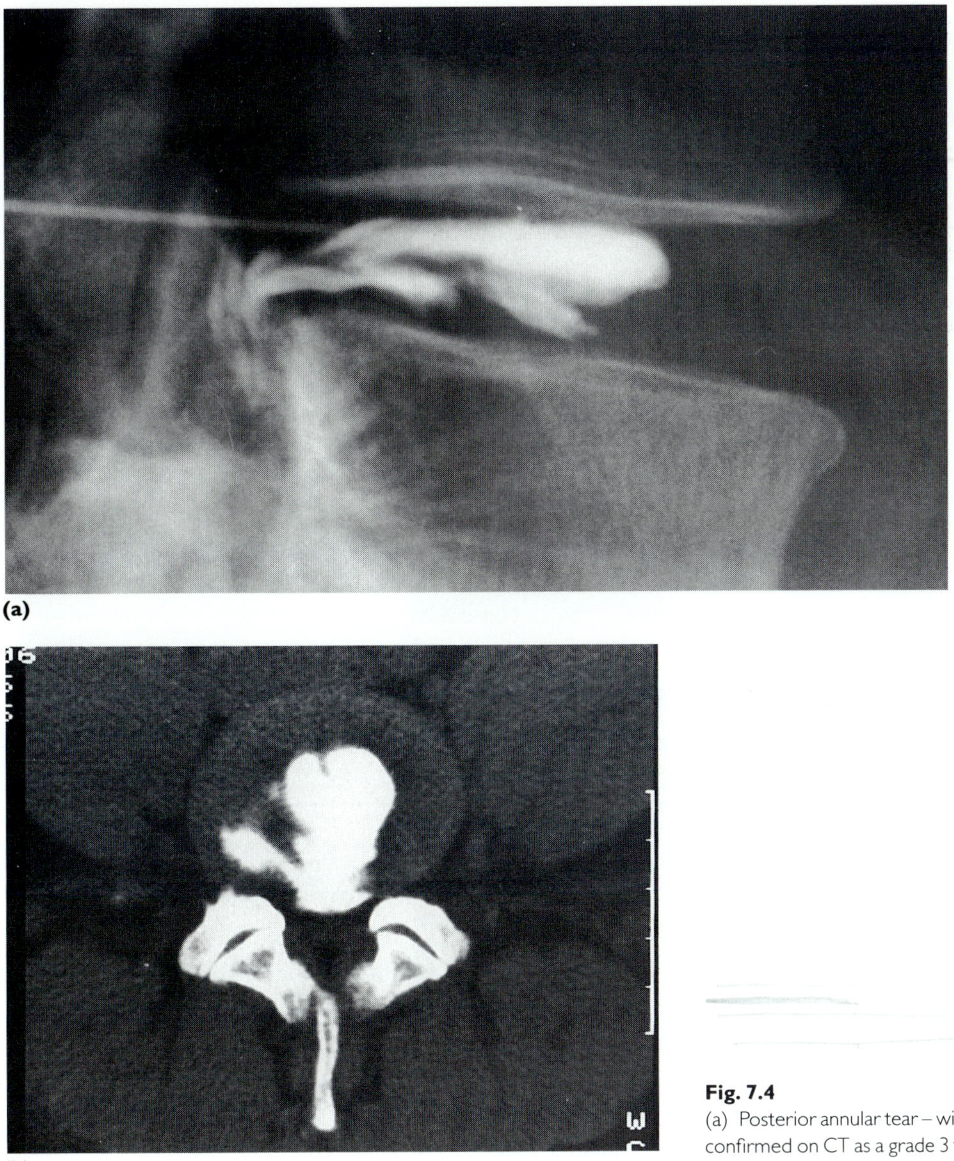

(a)

(b)

Fig. 7.4
(a) Posterior annular tear – with small extension of contrast medium – is confirmed on CT as a grade 3 tear (b).

2 A full annular tear is demonstrated on the plain radiographs by a line of contrast extending through the annulus and beyond the normal line of the annulus. The contrast may be contained by some intact outer peripheral fibres or the contrast may escape into the epidural space through the tear. The precise line of the tear is more clearly demonstrated on CT and would be graded as 3 for full annular penetration. Some concentric extension may also occur (Fig. 7.4a and b).

3 Prolapse of the nucleus pulposus may occur through a posterior annular tear and can be outlined by the contrast medium, which will demonstrate whether it is a protusion contained by residual fibres or an extrusion with contrast escaping around the nuclear material and entering the epidural space. Occasionally, a sequestrated migrated fragment may be outlined by contrast but this is unreliable.

4 A weakness in the vertebral endplate may lead to a localized collapse and subsequent interosseous disc herniation of nuclear material through the break. The most common and clinically insignificant lesion is the Schmorl's node, which is seen just posterior to the centre of the endplate. More severe forms, usually in the anterior third, which may be related to significant compression or flexion stress to the spine in adolescents, can cause back pain. Contrast injected into the nucleus will flow into the vertebral defect. Interosseous disc herniation may be difficult to recognize on CT discography without standard lateral views.

5 Degeneration of the disc is a slow, gradual process as the proteoglycans of the disc gradually reduce,

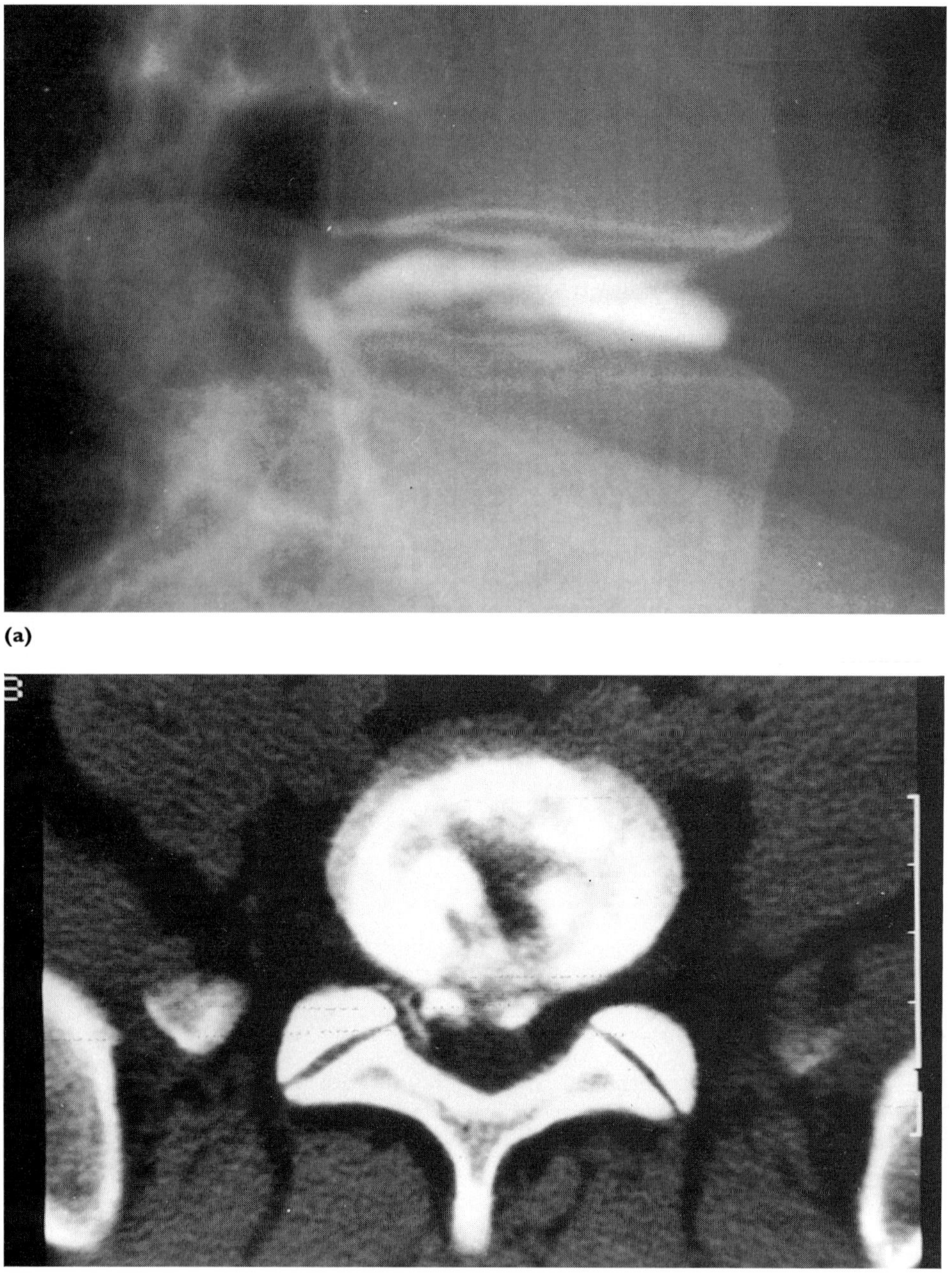

(a)

(b)

Fig. 7.5
(a) Degeneration of the disc is demonstrated, with contrast extending throughout the extent of the disc. This is confirmed on CT (b), which shows the central bulging of the annulus.

which reduces the water content. It may be part of the process of the preceding descriptions of disc rupture, but when it progresses it leads to narrowing of the disc space, and contrast injected into the disc extends throughout the disc on both the lateral and the AP view (Fig. 7.5a). The capacity of the disc is increased and it may take 3–4 ml of contrast. The posterior annulus may not be intact and contrast may leak into the epidural space. The CT grading is usually of the degenerative scale and may be 2 or 3 on the annular penetration assessment (Fig. 7.5b).

The demonstration of the anatomy of the nucleus pulposus and its surrounding annulus is of only limited value as disc degeneration and annular disruption are features of the normal ageing process. The structure of the nucleus and annulus is now also well demonstrated by means of MRI, and a good correlation has been shown between the MRI appearances and the discogram findings.[8,9] Cases where abnormalities have been shown on discography, in which the MRI results have judged to be normal, have however been reported.[10] Zucherman *et al.*[11] suggest that discography should be considered for patients with normal MRI and severe symptoms.

Pain production

The level of anatomical specificity of the disc appearance makes the assessment of stimulated pain a vital part of the investigation. Pain may be induced during needle insertion by the close proximity of the needle track to the lumbar nerve root exiting at the disc level being examined. A typical sharp pain or 'electric shock' in the dermatomal pattern of the nerve root will clearly identify this occurrence. Withdrawl of the needle and realignment will resolve this, and the pain is transient.

Needle placement against the annulus may induce pain. The outer annulus is innervated[12,13] and the pain induced is usually on the side of the needle and is relatively localized. If the disc is particularly sensitive, components of the symptomatic pain may be induced and this can be severe in some cases.

Finally, pain may be induced on injection of contrast medium into the nucleus. The use of non-ionic low-osmolar contrast agents with low neurotoxicity is recommended as they reduce the incident of spurious pain stimulation from non-discogenic sources in the event of leakage of the contrast into the epidural space.

The degree of pain production can be assessed by means of a scale of 0–10 (0 being no pain and 10 being the worst possible pain). The patient is also asked to state whether the pain produced is a pressure sensation, dissimilar from clinical pain, similar to the original pain or exactly reproducing the original pain.

Pain production from normal discs is rare but an experience of feeling pressure in the back is occasionally reported. Walsh *et al.*[14] showed that only one disc produced pain in a series of asymptomatic controls although a few false positives were reported by Vanharanta *et al.*[15] Symptom provocation from degenerate or injured discs is variable and, in the same study of discography of asymptomatic controls, no pain was induced from the abnormal discs.[14] In those subjects without a herniated nucleus pulposus, the posterior annular tears, which are grade 3, most commonly reproduce the patient's symptoms. Colhoun *et al.*[16] found that, in their series of patients with severe unresponsive back pain, 65 per cent of posterior annular tears without significant disc degeneration had reproduction of symptoms. Similar high levels of pain reproduction for grade 3 annular tears have been found by Vanharanta *et al.*[15] and Osti and Fraser.[10] Simple internal cleft formation in the posterior annulus, which would be grade 1 or 2, is least commonly painful, being found in 20 per cent of cases by Colhoun *et al.*[16] and 24 per cent by Vanharanta *et al.*,[15] but in these cases the pain is a deep-seated dull ache in the back with radiation often deep inside the limbs.[7] Symptomatic pain in patients with interosseous disc herniations was reported by McCall *et al.*[17] in young patients with a history of trauma. This lesion is not common, being present in only 14 of 692 discs injected by Hsu *et al.*,[18] but concordant symptoms were also induced in 13 out of the 14 by these authors. This high incidence of symptom reproduction was not found in a small series of older chronic back pain sufferers.[16]

The precise relationship between severe disc degeneration and pain is not clear in view of the fact that disc degeneration is very common with age and that it frequently exist without pain.[19] It remains an issue of interest why some degenerative discs are sympto-matic and others are not. Colhoun *et al.*[16] found that degenerative discs were found to be the cause of symptoms in only 50 per cent of patients with severe back pain. Although degenerated discs were symptomatic, in two studies by Vanharanta *et al.*[15, 20] they varied in frequency between 64 and 80 per cent. Fraser *et al.*,[21] in an analysis of 370 discographically proven degenerate discs, found that symptom reproduction occurred most commonly when the annulus bulged beyond the posterior vertebral line and the contrast was contained as compared to those where the annulus did not bulge or where the contrast extravasated extensively into the epidural space. This comparison was statistically significant at the 0.1 per cent level. Recent studies by Maezawa and Muro[22] comparing the CT discographic appearance and symptom reproduction also showed that a grade 3 annulus disruption, with a focal disc protusion beyond the line of the vertebral body, had a significantly higher frequency of symptomatic reproduction for all three grades of degeneration as compared to those without disc protusion. Back pain, when part of the symptom pattern, is regularly reproduced by the injection whilst buttock and thigh pain, when present, is reproduced in 80 per cent of cases. Pain below the knee at presentation is rarely reproduced at discography.[16]

Following the stimulation of pain by contrast medium, the pain may be blocked by injecting a 0.5 per cent bupivacaine hydrochloride into the disc. Partial or complete relief of the induced pain may occur but the exact relationship of pain relief to outcome of treatment has not been reported. If there is significant escape of contrast into the epidural space, the injection of bupivacaine is not advised as it may anaesthetize other l evels, causing inaccuracies in stimulation on subsequent discography at these levels. The immediate relief of symptoms does, however, act to support the diagnostic examination.

Delayed onset of symptoms has been reported in patients who did not have an immediate pain response to the injection. It has been described in discs with incomplete or discrete annular tears, which have not been clearly demonstrated at the time of the investigation but were seen 2–12 hours after the examination.[23] Review of patients at 24 hours following the examination may be advisable although the precise level of symptoms may be difficult to assess if a three-level examination has been performed.

Predictive value of lumbar discography

The accuracy of discography in identifying stages of disc degeneration is not doubted.[5] The fact that pain, which may be similar to the patient's symptoms, may also be provoked by injection of contrast is also accepted. Dispute remains, however, as to how accurate this pain provocation is in relation to individual patients.[24]

Holt,[25] in a study on pain-free subjects, found that severe backache after disc injection was produced in 15 per cent and combined backache and sciatica in 22 per cent. However, the contrast medium used was sodium based, leaked into the epidural space in these subjects and was probably responsible for the pain production. Walsh *et al.*,[14] in a carefully controlled discographic study on 10 asymptomatic volunteers and 7 patients with back pain, produced no significant pain on injection with non-ionic contrast medium in the asymptomatic group, a specificity of 100 per cent.

The diagnostic accuracy of discography in patients with back, hip or leg pain, without disc herniation and nerve root compression, is difficult to assess as the end-point against which to judge its accuracy is surgery, which may vary in type and success.

Simmons and Segil[26] compared the value of clinical assessment, myelography and discography in patients who had undergone either discectomy alone, discectomy with fusion or fusion alone. Analysis was based on relief or significant improvement of preoperative symptoms by operation at the selected level or levels. Their results indicated that discography was accurate in 82.2 per cent of cases, whereas clinical examinations and myelography had accuracies of 44.2 per cent and 45.6 per cent respectively. Brodsky and Binder[27] found that 69.4 per cent of discograms fully matched the patients' pain and 13.6 per cent had partial replication. In 78 per cent of cases, discography had significantly influenced operative decisions.

Colhoun *et al.*[16] studied a homogeneous and uncomplicated group of 189 patients in whom a technically successful fusion had been performed and found that measured against a clinically successful outcome of surgery, discography was accurate in 82 per cent of cases. The sensitivity was high at 89 per cent but the specificity was only 43 per cent, as 52 per cent of patients without pain reproduction on discography also had a good result from surgery. The lower specificity may in part be explained by the fact that fusion may also resolve pain produced from other components of the motor segment. Of particular interest in this study, however, was the fact that when a morphologically abnormal but painless disc was left untouched at the time of fusion of the painful level, the clinical outcome was good in 95 per cent of cases. Findings contrary to those of Colhoun *et al.*[16] have been reported by Esses *et al.*[25] who found in a series of 35 patients that temporary external spinal fixation was more accurate in predicting the outcome of fusion than discography, which showed no statistical correlation between the eventual surgical outcome and symptom reproduction at discography. However, this study had a high proportion of patients who had undergone previous low back surgery and multilevel fusion was the norm.

In patients with disc herniation, Hudgins[29] found an accuracy of pain reproduction of between 75 and 85 per cent, whereas Jackson *et al.*[2] using CT discography, found that only 26 per cent of 125 proven nucleus pulposus herniations produced typical sciatic pain reproduction. However, Vanharanta *et al.*,[20] also using CT discography, found that 80 per cent of patients with disc herniation had pain reproduction.

Some of this discrepancy may be explained by the limited level of peripheral pain reproduction found by Colhoun *et al.*[16] and also by the fact that pain reproduction for discogenic pain follows a different mechanism as compared with sciatica, the latter being due to nerve compression.

The basis of the symptomatic pain reproduction at lumbar discography is not clearly understood. The presence of nerve endings in the annulus and adjacent longitudinal ligaments is not in dispute,[12,13] and recent work on labelling for substance P has also confirmed that these are pain-related nerve endings.[30] There may be a direct relationship to increased pressure from the injection with stretching of the outer annulus and pain production.[21,22] Accumulation of solutes which are depolarizing agents may produce chemical irritation of the nociceptors system, which may also give rise to pain.[31] These authors suggest that concomitant pain, sometimes associated with an abnormal discogram image, may in part be related to the chemical environment within the intervertebral disc and the sensitized state of its 'annular nociceptors'.

Complications of lumbar discography

Lumbar discography is an essentially safe procedure provided that care is taken with the technical positioning of the needle. If the approach is too lateral the needle may pass outside the paraspinal musculature and transgress the peritoneum, with possible bowel penetration.

Nerve root injury may occur as the needle passes across the exit zone of the nerve root canal. It is essential that the patient is fully conscious so that needle irritation of the nerve will be felt by the patient as a sharp pain in the root distribution. The needle should be withdrawn and repositioned and no long-term sequelae should occur.

The main complication is post-discography disc

infection. This complication has been considered to be rare,[26, 32,33] and it has been attributed in part to a chemical or aseptic reaction[27,34,35] by some authors. Guyer *et al.*[36] found an incident of 0.1 per cent of patients and 0.05 per cent of discs investigated. Fraser *et al.*,[37] however, reviewed their experiences with discography and found a high incidence of infection (2.7 per cent) when using a large, 18 G needle, whereas the rate fell to 0.7 per cent when stiletted two needle technique was used. Discitis is diagnosed when a patient has a marked increase or acute sustained pain in the back and a raised erythrocyte sedimentation rate or C-reactive protein following discography. Plain films are often unchanged for a number of weeks but early imaging changes are provided by increased activity on ^{99}Tc Methylene Diphosphonate (MDP) bone scanning, although Guyer *et al.* did not find positive bone scans in under 18 days following the investigation.[36] Diagnosis is also demonstrated early by MRI, which shows decreased signal on T1 and increased signal on T2 in the endplate, with enhancement of the T1 weighted image after injection of gadolinium. Erosion of the endplate on the plain films or tomograms will occur after about 3 weeks.

A positive culture may be achieved in only 50 per cent of patients with discitis although evidence of an infective process was found on histology in all suspected cases undergoing subsequent anterior surgery.[37] Experimental work with sheep confirms that discitis is an infective rather than a chemical phenomenon. Varied injected dilutions of bacterial suspension injected into the discs have demonstrated typical radiographic changes of discitis but the controls using only contrast media did not. Culture from the discs only revealed an infection in the first 2 weeks following inoculation but histology confirmed the infective change from week 2. The reaction produced by a single organism was the same as that produced by a heavy inoculation but in neither could the organism be isolated after 6 weeks.[37] Because of its isolation from vascular tissue, the normal nucleus pulposus may be considered an ideal culture medium. The initial histological changes are granulation tissue on the vertebral side of the endplate with small defects in the hyaline cartilage. Once the endplate is breached the bacteria are rapidly removed.

The risk of infection can be reduced by carrying out the intradiscal injection in a clean room with full aseptic techniques. The skin should be prepared over a wide area to allow for patient movement. Care should be taken to avoid contamination by radiographic equipment. The two-needle technique with 21 and 26 G needles is recommended and stilettes should be used to avoid implantation of skin in the disc. The operator should avoid touching the needle shaft and tip, and separate needles should be used for each disc level.

CERVICAL DISCOGRAPHY

The first use of cervical discography has been attributed to Smith and Nichols,[38] but it was pioneered by Cloward,[39] who emphasized pain reproduction as the key feature of the examination. As with lumbar discography, it is an invasive investigation but it also carries a higher risk of complication. The advent of CT and MRI has reduced the indications and it is now of value only for those cases of cervical discogenic pain whose symptoms are severe enough to warrant surgical intervention. It is not indicated if the symptoms, signs or electromyographic (EMG) studies indicate that a radiculopathy is present. In such cases, cervical myelography, CT myelography or MRI are the investigation techniques of choice. The discogenic neck pain is sclerotomal, with a dull and aching quality, and is perceived in the central neck and both lateral aspects of the neck base, over the shoulders and between the shoulder blades.

PREPARATION AND INFORMED CONSENT

The patient will require a detailed explanation of the procedure and should be clearly informed of the potential complication of infection. The patient must be advised that they may feel pain and that they should inform the radiologist where their pain is being felt. The patient should be carefully assessed to exclude any possible source of infection and the procedure postponed if one is present. Evidence of spinal cord compression constitutes an absolute contraindication of discography. In such cases, information as to whether a disc is painful or not is immaterial.

Mild sedation with diazepam (10 mg, 1 hour prior to the investigation) may be helpful. Intravenous broad-spectrum antibiotics may be of value just prior to the investigation, although the efficacy of this procedure has not been established in cervical discography.

The examination is undertaken with the patient supine and the neck extended over a soft pad although hyperextension should be avoided. Biplanar screening is essential and this can be achieved with a C-arm screening unit either on its own or in combination with a standard screening table. An overhead tube screening system is an advantage as it allows easy access to the patient's neck. This is especially useful in cervical discography as, unlike in lumbar discography, needle movement during placement is common because of swallowing, and also because of limited support of the needle by the surrounding soft tissues.

Careful skin preparation and cleaning is essential, and repeated cleansing with antiseptic solution is required. The operator should be fully gowned and careful draping of the patient's neck should be undertaken. The disc is approached from the antero-lateral aspect, medial to the carotid artery and lateral to the pharynx. A right-sided approach is favoured because the oesophagus lies to the left of the mid-line in the lower neck and should be avoided. The gap between the two structures is widened by gentle digital palpation and the level of injection is checked on AP screening.

Local anaesthetic is injected into the skin and down to the antero-lateral aspect of the vertebral body. A 21 G 3½ in. (90-mm) needle is inserted into the neck, medial to the medial border of the sternocleido-mastoid and not through the muscle, while again applying digital pressure to the neck to separate the pharynx medially and the carotid artery laterally. In the upper levels, the needle should be inserted more laterally to avoid puncturing the pharynx. The needle is advanced at a 45° angle down to the antero-lateral aspect of the annulus. Upon contact with the annulus, the patient may experience transient pain. The position is checked on the AP fluoroscopy. A 26 G needle is passed through the 21 G under careful lateral screening control so that the needle does not penetrate beyond the centre of the nucleus (Fig. 7.6). The patient is then screened in the AP view to check that the needle is in the centre of the disc. It is essential that the needle position should be accurate as the nucleus of the cervical disc is small.

INJECTION

Low-osmolar non-ionic contrast medium is used for the injection and is inserted with a 1-ml syringe. The normal nucleus will take only approximately 0.2–0.3 ml with marked resistance developing, and the contrast medium restricted to the centre of the disc on the AP and lateral view.

Completely normal discs are uncommon after the mid-twenties and the more common picture is of

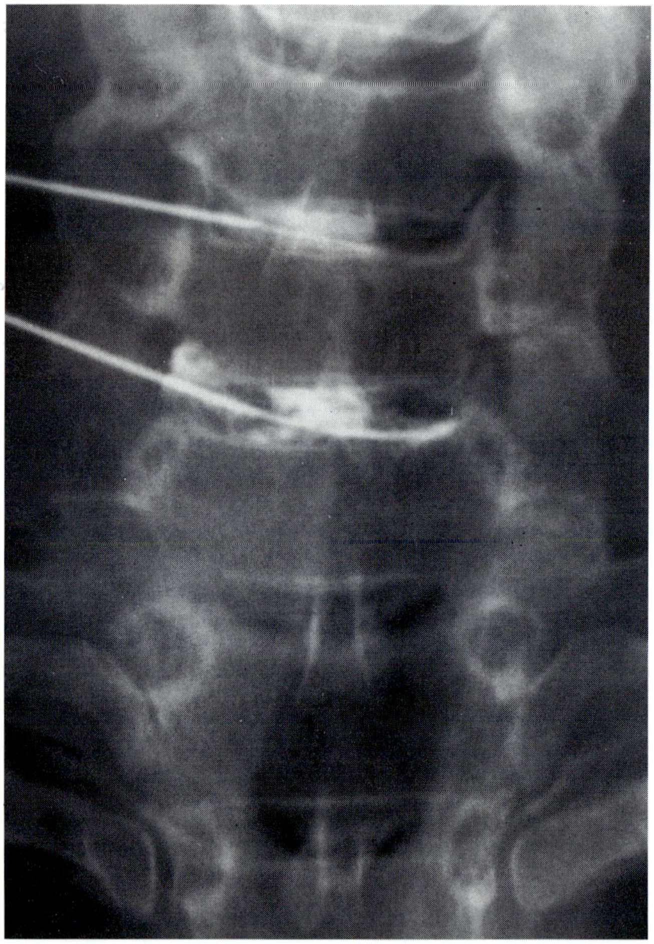

Fig. 7.6
The 21 G needle is positioned on the edge of the annulus antero-laterally. The 26 G needle is passed through it into the centre of the nucleus pulposus. Contrast medium injected is shown entering the neurocentral joints on the AP projection.

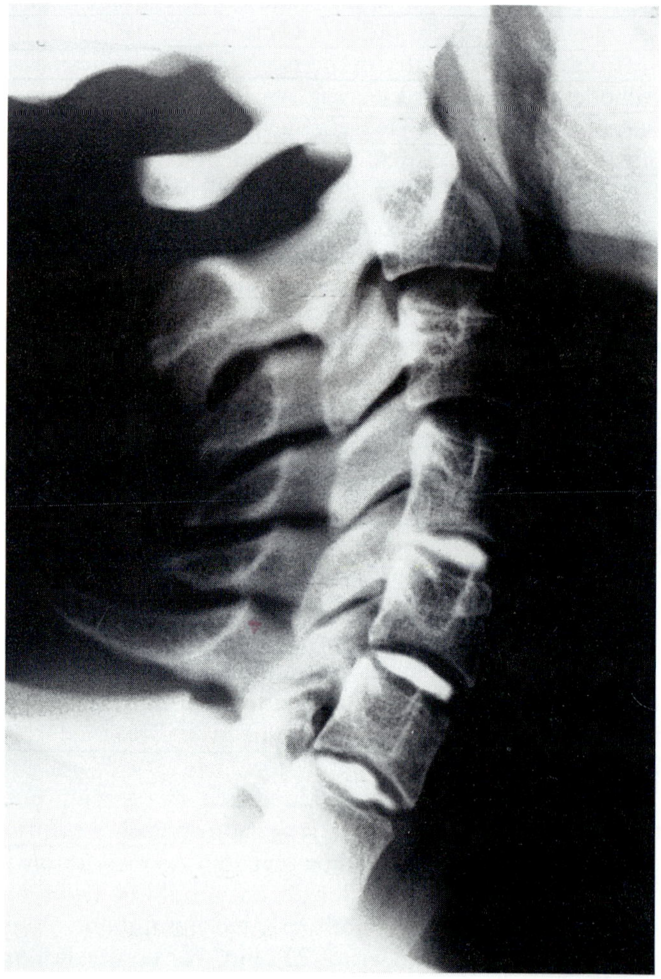

Fig .7.7
Three-level cervical discography has been perfomed. The C5/6 and C6/7 discs are normal. At C4/5 there is a posterior annular disruption.

contrast extending out through postero-lateral clefts into the neurocentral joints.[40] This is best appreciated on the AP view, and both an AP and a lateral X-ray film are taken with the needle *in situ*. The lateral view is analysed for evidence of a bulging disc or of a significant disc prolapse (Fig. 7.7).

The relationship between symptoms and the disc appearances is less clear in cervical discography than in lumbar discography. Any pain induced during the injection of contrast is recorded and the distribution, character and intensity of the pain is compared with the patient's normal symptoms. Roth[47] introduced the concept of analgesic discography in that the induced pain should be relieved by injecting local anaesthetic into the disc. Following the stimulation of pain by contrast, a small quantity of local anaesthetic (usually 0.1–0.2 ml) may be injected into the disc to assess any reduction in pain. Care must be taken not to use too much local anaesthetic if there is considerable epidural leakage through the posterior annulus as this may block a pain response from a subsequent disc injection. It is very unusual to inject more than three levels and it may be valuable to undertake the individual levels on different days if significant symptoms are induced.

Predictive value of cervical discography

The innervation of the cervical disc is complex and nerve endings have been demonstrated. The cervical disc is innervated by the sinu-vertebral nerves but also laterally from the vertebral nerve and the cervical sympathetic trunk.[42]

The accuracy of cervical discography is difficult to assess. There is more overlap of the sclerotomal zones in the cervical spine as compared with the lumbar spine but Simmons and Segil[26] found a good correlation between the demonstration of symptoms and the result of subsequent surgery in 91 per cent of patients, whilst other authors have suggested that 70 per cent have proved to have good results following demonstration of a symptomatic level.[43] Kikuchi *et al.*[44] localized symptomatic discs by discography and by establishing that the discogenic pain could be relieved by appropriate nerve blocks, and found that their operation results improved, At 1-year follow-up, 80 per cent of their patients were pain free compared to 40 per cent when discography was not used. The investigation is not without its detractors, however,[45,46] who conclude that it is non-specific in both anatomical demonstration and symptom reproduction. In a recent small study, Connor and Darden[47] found a positive predictive value of a good result of surgery in only 46 per cent of cases.

Complications

Misplacement of the needle and infection are the main complications of cervical discography. Neural structures are not at risk during the course of the needle to the target disc but the short sagittal diameter of the cervical disc may allow the passing of the needle into the cord. It is therefore essential to screen the inner needle during its insertion into the centre of the disc.

Penetration of viscera such as the pharynx and oesophagus is not a problem in itself but it does increases the risk of infection. Pneumothorax is a potential risk at C7/T1. Arterial puncture is a hazard at all levels but the use of fine needles and attention to proper technique, with good visualization, should avoid the occurrence of such complications. Cervical discography is not without hazard as subsequent disc infection may occur, and the incidence appears significantly higher than the 0.7 per cent quoted for lumbar discography.[3] Guyer *et al.*[36] found that 2 out of 145 patients developed discitis following cervical discography (1.3 per cent). The incidence of discitis per disc level was 0.5 per cent (2 of 362 discs). Clinically there is a marked exacerbation of pain and disability and this is accompanied by an elevated sedimentation rate. The bone scan may demonstrate increased uptake but is not usually positive for 2 weeks. Disc space narrowing, pre-vertebral soft tissue widening and endplate destruction occur initially, followed at a large stage by fusion. Severe complications in the form of epidural or subdural abscess have been reported in three cases which subsequently progressed to quadraplegia.[47–49]

CONCLUSION

The role of discography is constantly being re-evaluated. The advent of magnetic resonance imaging, which can reveal the nature of disc structure and damage, has focused attention again on discography. The finding that many discs that are abnormal on MRI are not necessarily painful and that occasionally symptomatic discs are normal on MRI lends support to the continued requirement for provocative disc injections. MRI provides a useful screening procedure prior to discography, particularly in the cervical spine, to diagnose significant disc herniation and extensive disc degeneration. The presence of a single abnormal level may be helpful but, in the lumbar spine, discography should still be performed if fusion is contemplated. Cervical discography should be undertaken only when fusion is to be performed and the patient should be warned of the possible complications.

References

1 Lindblom K (1948). Diagnostic disc puncture of intervertebral discs in sciatica. *Acta Orthopaedica Scandinavica* **17**: 231–9.
2 Jackson RP, Cain JE, Jacobs PR, Cooper BR, McClann GE (1989). The neuroradiographic diagnosis of lumbar herniated nuclear pulposus. *Spine* **14**: 1356–61.
3 Fraser RD, Osti OL, Vernon Roberts B (1989). Iatrogenic discitis: The role of intravenous antibiotics in prevention and treatment: An experimental study. *Spine* **14**: 1025–31.
4 Quinnell RC, Stockdale HR (1980). An investigation of artefacts in lumbar discography. *British Journal of Radiology* **53**: 831–9.
5 Adams MA, Dolan P, Hutton WC (1986). The stages of disc degeneration as revealed by discograms. *Journal of Bone and Joint Surgery* **68B**: 36–41.
6 Sachs B, Vanharanta H, Spivey MA, Guyer RD, Videman T, Rashbaum RF (1987). Dallas discogram description: A new classification of CT/discography in low back disorders. *Spine* **12**: 287–94.
7 Crock HV (1970). A re-appraisal of the intervertebral disc lesions. *Medical Journal of Australasia* **1**: 983.
8 Collins JP, Stack JP, O'Connell DJ, Walsh M, McManus HP, Redmond OM, Ennis JT (1990). The role of discography in lumbar disc disease: Comparative study of MRI and discography. *Clinical Radiology* **42**: 252–7.
9 Schneiderman G, Flannigan B, Kinsgston S *et al.* (1987). Magnetic resonance imaging in the diagnosis of disc degeneration correlation with discography. *Spine* **12**: 276–81.
10 Osti OL, Fraser RD (1992). MRI and discography of annular tears and intervertebral disc degeneration. *Journal of Bone and Joint Surgery* **74B**: 431–5.
11 Zucherman J, Derby R, Hsu K *et al.* (1988). Normal magnetic resonance imaging with abnormal discography. *Spine* **13**: 1355–9.
12 Yoshizawa H, O'Brien JP, Smith WT, Trumper M (1980). The neuropathology of intervertebral discs removed for low back pain. *Journal of Pathology* **132**: 95–104.
13 Bogduk N, Tynan W, Wilson AS (1981). The nerve supply to the human lumbar intervertebral discs. *Journal of Anatomy* **132**: 39–56.
14 Walsh TR, Weinstein JN, Spratt KF, Lehmann TR, Aprill C, Sayre Hulha (1990). Lumbar discography in normal subjects. *Journal of Bone and Joint Surgery* **72A**: 1081–8.
15 Vanharanta H, Sachs BL, Spivey MA, Guyer RD, Hochschuler SH, Rashbaum RF (1987). The relationship of pain provocation to lumbar disc degeneration as seen by CT/discography. *Spine* **12**: 295–8.
16 Colhoun E, McCall IW, Williams W, Cassar-Pullicino VN (1988). Provocative discography as a guide to planning operations on the spine. *Journal of Bone and Joint Surgery* **70B**: 267–71.
17 McCall IW, Park WM, O'Brien JP, Seal PV (1985). Acute traumatic interosseous disc herniation. *Spine* **10**: 134–7.
18 Hsu K, Zuckerman JF, Derby R, White AH, Goldwaite NG, Wynne G (1988). Painful lumbar end plate disruptions. A significant diagnostic finding. *Spine* **13**: 76–8.
19 Powell MC, Wilson A, Szyprt P, Symonds EM (1986). Prevalence of lumbar disc degeneration observed by magnetic resonance in symptomatic women. Lancet **ii**: 1366–7.
20 Vanharanta H, Guyer RD, Ohnmeiss DD, Stilh WJ, Sachs BL, Aprill C (1988). Disc deterioration in low back syndromes: A prospective multicentre CT/discography study. *Spine* **13**: 1349–51.
21 Fraser J, McCal IW, Park WM, O'Brien JP (1983). Discography in degenerative disc disease. *Orthopaedic Transactions* **7**: 466.
22 Maezawa S, Muro T (1992). Pain provocation at lumbar discography as analysed by computed tomography/discography. *Spine* **17**: 1309–15.
23 Lehmer SM, Dawson MHO, O'Brien JP (1994). Delayed pain response after lumbar discography. *European Spine Journal* **3**: 28–31.
24 Nachemson A (1989). Editorial comment. Lumbar discography – where are we today? *Spine* **14**: 555–7.
25 Holt AP (1968). The question of lumbar discography. *Journal of Bone and Joint Surgery* **50A**: 720–5.
26 Simmons EH, Segil CM (1975). An evaluation of discography in the localisation of symptomatic levels in discogenic disease of the spine. *Clinical Orthopaedics* **108**: 57–69.
27 Brodsky AE, Binder WF (1979). Lumbar discography: Its value in diagnosis and treatment of lumbar disc lesions. *Spine* **4**: 110–20.
28 Esses SI, Botsford DJ, Kostuk JP (1989). The role of external spinal fixation in the assessment of low back disorders. *Spine* **14**: 594–600.
29 Hudgins WR (1977). The diagnostic accuracy of lumbar discography. *Spine* **2**: 305–9.
30 Ashton IK, Roberts S, Jaffray DC, Polak JM, Eisenstein SM (1993). Neuropeptides in the human intervertebral disc. *Journal of Orthopaedic Research* **12**: 186–92.
31 Weinstein J, Claverie W, Gibson S (1988). The pain of discography. *Spine* **13**: 1344–8.
32 Collins JS, Gardener WJ (1962). Lumbar discography: An analysis of one thousand cases. *Journal of Neurosurgery* **19**: 452–61.
33 McCulloch JA, Waddell G (1978). Lateral lumbar discography. *British Journal of Radiology* **51**: 498–502.
34 McCulloch JA (1977). Chemonucleolysis. *Journal of Bone and Joint Surgery (British volume)* **59B**: 45–52.
35 Crock HV (1983). *Practice of Spinal Surgery*. New York Springer-Verlag.
36 Guyer RD, Collier R, Stith WJ, Ohnmeiss DD, Mochschuler SH, Rashbaum RF, Regan JJ (1988). Discitis after discography. *Spine* **13**: 1352–4.
37 Fraser RD, Osti OL, Vernon Roberts B (1987). Discitis after discography. *Journal of Bone and Joint Surgery* **69B**: 26–35.
38 Smith GW, Nichols P (1957). The technique of cervical discography. *Radiology* **68**: 718–20.
39 Cloward RB (1958). Cervical discography technique: Indications and use in diagnosis of ruptured cervical discs. *American Journal of Roentgenology* **79**: 565–74.
40 Sneider SE, Winslow OP, Pryor JH (1963). Cervical discography: Is it relevant? *Journal of the American Medical Association* **185**: 163–5.

41 Roth DA (1976). Cervical analgesic discography: A new test for the definitive diagnosis of the painful disc syndrome. *Journal of the American Medical Association* **235**: 1713–14.

42 Bogduk N, Windsor M, Inglis A (1989). The innervation of the cervical intervertebral discs. *Spine* **13**: 2–8.

43 Whitecloud TS, Seago RA (1987). Cervical discogenic syndrome: Results of operative intervention in patients with positive discography. *Spine* **12**: 313–16.

44 Kikuchi S, McNab I, Moreau P (1981). Localisation of the level of symptomatic cervical disc degeneration. *Journal of Bone and Joint Surgery* **63B**: 272–7.

45 Klafta LA, Collins JS (1969). The diagnostic inaccuracy of the pain response in cervical discography. *Cleveland Clinical Quarterly* **36**: 35–9.

46 Meriam WF, Stockdale HR (1983). Is cervical discography of any use? *European Journal of Radiology* **3**: 138–41.

47 Connor PM, Darden BV (1993). Cervical discography: Complications and clinical efficacy. *Spine* **18**: 2035–8.

48 Eismont FJ, Bohlmann HH, Prasanne L *et al.* (1983). Pyogenic and fungal vertebral osteomyelitis, with paralysis. *Journal of Bone and Joint Surgery* **65A**: 19–29.

49 Lownie SP, Ferguson GG (1989). Spinal subdural empyema complicating discography. *Spine* **14**: 1415–17.

CHAPTER 8

Chemonucleolysis

John A. McCulloch

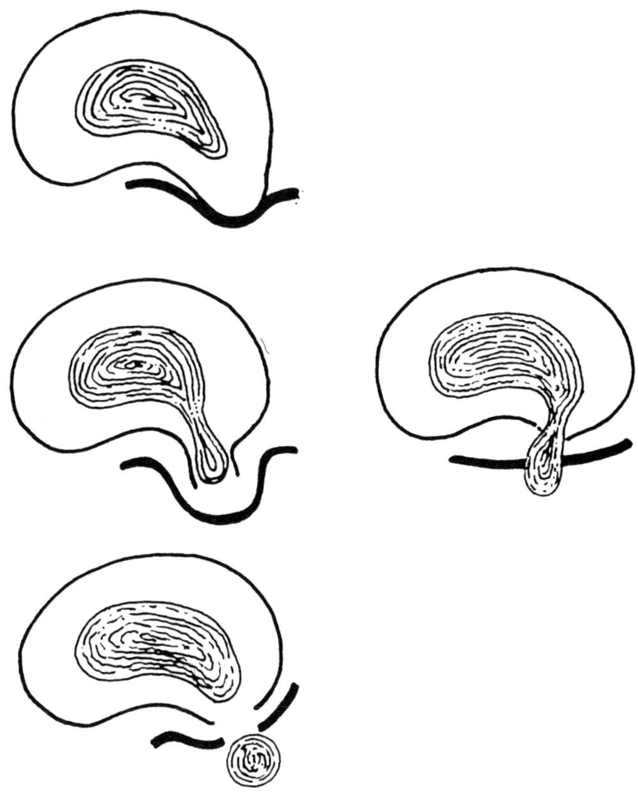

INTRODUCTION

Surgery for degenerative lumbar spine problems is quickly diverging in two directions: (i) least invasive and (ii) most invasive procedures. The middle ground of standard laminectomy and discectomy shrinks as percutaneous and microsurgical procedures become more popular, and pedicle screws, laminar hooks, plates and rods take us to more aggressive levels of surgical intervention. Helping to forge these new directions are a much better understanding of clinical syndromes, and new diagnostic methods such as MRI. Early intervention with conservative, non-surgical treatment, based on clear diagnoses, has a profound effect on lumbar disc disease.

Two seminal events stand out in the understanding of degenerative conditions of the lumbar spine: Mixter and Barr's description of the lumbar disc rupture[1] and Verbiest's description of spinal stenosis.[2] The most dramatic recent breakthrough in understanding the lumbar spine has been the advent of MRI, which has become the eyes of the spine surgeon intent on considering least invasive procedures for lumbar disc ruptures.

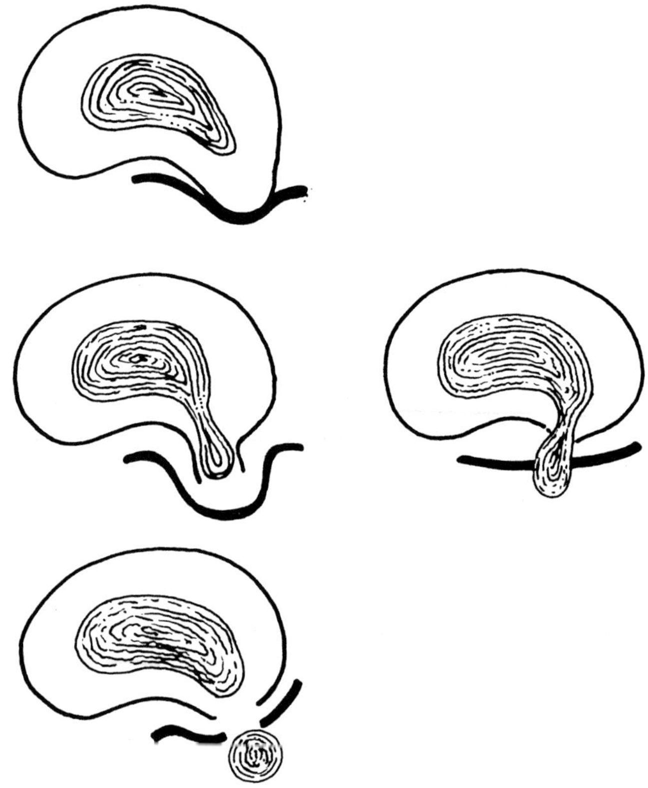

Fig. 8.1
Contained and non-contained disc ruptures. Top, a contained protusion; middle, two types of disc extrusions: left, a contained subannular extrusion; right, a non-contained, transannular disc extrusion; bottom, a sequestered disc.

CLASSIFICATION

For the moment, let us set aside the more open, multilevel procedures that are associated with spinal instrumentation and let us consider the matter of a single-level disc prolapse or herniation causing sciatica.

There are two basic types of disc ruptures (Fig. 8.1): contained and non-contained. Each represents a focal rupture, prolapse or herniation of intradiscal substance (usually nucleus pulposus) beyond the confines of the annulus into nerve root territory. Anterior disc ruptures (Fig. 8.2): are asymptomatic. It is only, but not always, when discal material irritates and/or compresses nerve roots that radicular pain (sciatica) travels down the leg.

Contained disc herniations may be protrusions or subannular-ligamentous extrusions. Since the posterior longitudinal ligament in the lower lumbar spine is so insignificant, these herniations will be described in relation to their relationship with the annulus. The non-contained disc herniations include transannular herniations (very common), and sequestered discs (very rare) in which the fragment of ruptured disc lies free in the spinal canal with no connection to the disc space. On a few occasions, a sequestered disc's origin cannot be discovered in spite of extensive inspection, but most often the rent in the annulus through which the sequestered fragment of disc ruptures is evident.

THE SYMPTOM

Twenty-five per cent of disc ruptures seen on MRI have the potential for being asymptomatic.[3] When a disc rupture causes symptoms, they take the form of sciatica (Fig. 8.3). When a degenerative disc, without rupture, causes symptoms, they are in the form of back pain. Disc ruptures cause different clinical presentations in different age groups (Table 8.1).

INDICATIONS FOR SURGICAL INTERVENTION

Table 8.2 outlines the indications for surgical intervention. The *absolute* indications for some form of intervention represent the contraindications for chemonucleolysis. Chemonucleolysis, as interventional therapy, is indicated only in those situations that are outlined in Table 8.2 as relative indications. It has been proposed that the use of chymopapain be considered as the last step in conservative care rather than the first step in surgical intervention.

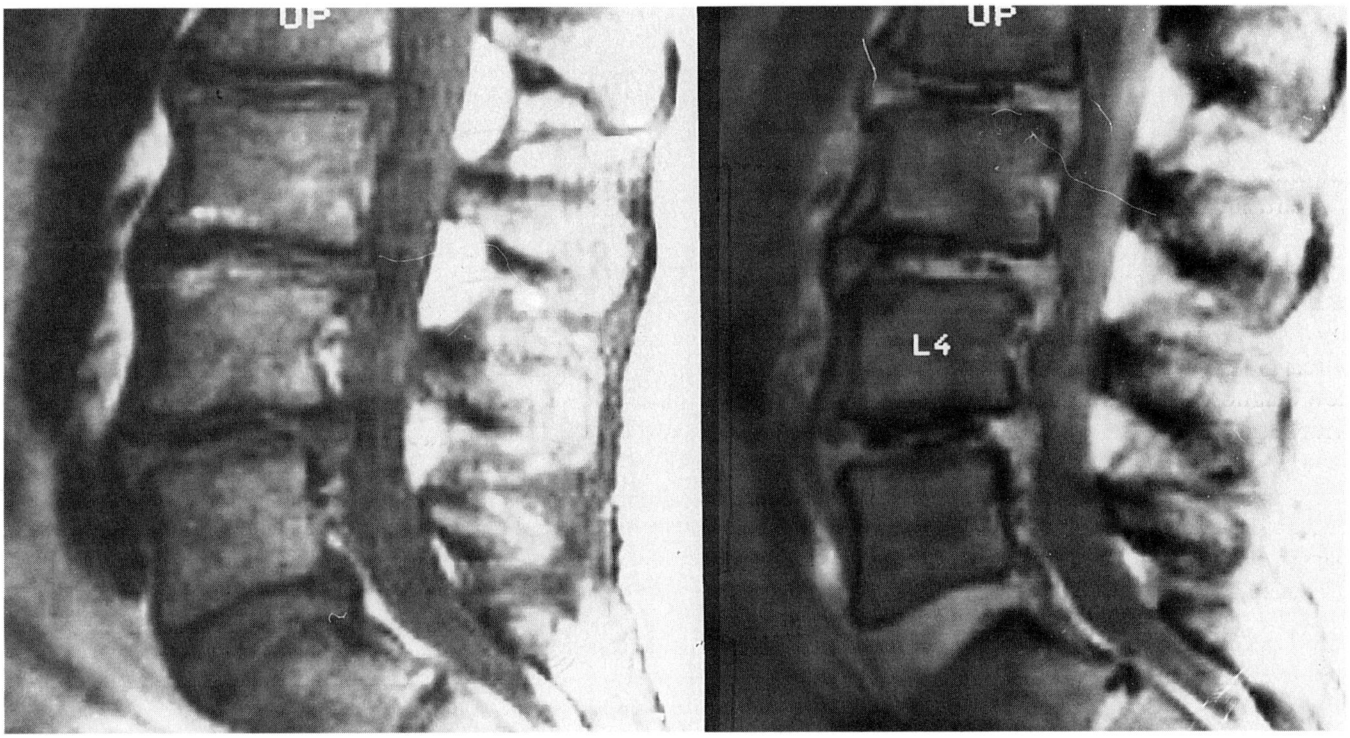

Fig. 8.2
Multiple MRI slices showing anterior disc herniation at L4–5. The symptomatic lesion was the posterior mid-line focal HNP at L4–5.

	Age		
Symptom	Adolescent (<25 yrs)	Adult (30–50)	Senior adult (55–80)
Pain	Typical radicular pattern, may not be below knee	Typical radicular pattern, almost always below knee	Typical radicular pattern, most severe below the knee
Paresthesiae	50% chance of being present	Common	Most common
SLR reduction	Profound	Less than 50% of normal	Most often more than 50% of normal
Neurological signs	>50% chance of being absent	>50% chance of being present	Most often present
Associated degenerative changes (spinal stenosis)	Rare	Occasional	Common
Response to conservative care	Recurrence rate of symptoms very high	Good response to conservative care	Limited tolerance for prolonged care
Protrusion/ extrusion	Protrusions very common	Protrusions less common	Protrusions rare

Table 8.1 *Clinical picture of sciatica in different age groups*

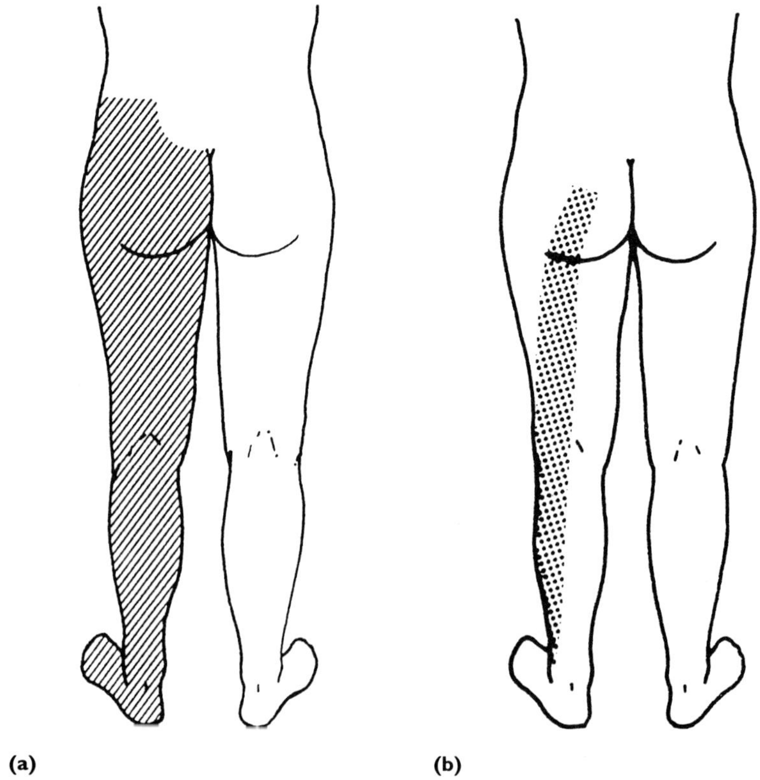

Fig. 8.3
(a) The proximal portion of the leg includes the buttock. (b) Radicular pain follows a nerve root pattern in its radiation.

Absolute
1. Cauda equina syndrome
2. Increasing neurological deficit
3. Significant neurological deficit with large disc fragment on MRI
Relative
1. Failed conservative treatment
2. Recurrent sciatica

Table 8.2 *Indications for interventional therapy in lumbar disc prolapse*

Chymopapain

Since Smith[4] proposed the use of chymopapain for dissolving disc herniations, it has followed a chequered course to clinical acceptance. Widely used in Europe and Asia, it is almost extinct as a form of treatment intervention in North America. In Canada, the procedure is limited in use because of very low fees/risk ratio set by the government-run health care system; in the United States the drug is not used because of the potential for neurological and allergic complications that may occur in a very sour malpractice environment.

The author's personal experience with chymopapain is approaching 8000 patients over 24 years without a single serious permanent complication. During that time 19 anaphylactic reactions have occurred[5] as well as 1 temporary subarachnoid haemorrhage ending in full recovery. One can only conclude, after this experience, that the drug is inherently safe unless placed in spaces unintended, such as the subarachnoid space.

Action of chymopapain

Chymopapain is an extract of latex of the tropical fruit papaya[6]. Of the proteolytic enzymes in papaya, chymopapain is the most specific in its activity on nucleus pulposus and the least antigenic. Despite its being less antigenic than papain, chymopapain is a protein foreign to the human body and capable of precipitating allergic reactions, including anaphylaxis.

Connective tissue can be classified into two basic components: cells and non-cellular matrix. The cells are cartilage connective tissue cells and located in the nucleus pulposus; the matrix is made up of collagen and proteoglycan–water complexes. This non-cellular matrix

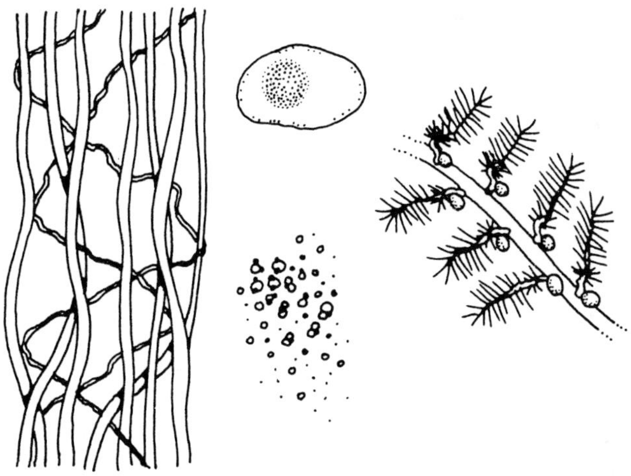

Fig. 8.4
The make-up of the nucleus. To the left, collagen; to the right, proteoglycan; middle top, the cartilage cell; middle bottom, water. Chymopapain splits the side-chains of proteoglycan and interferes with its ability to hold water.

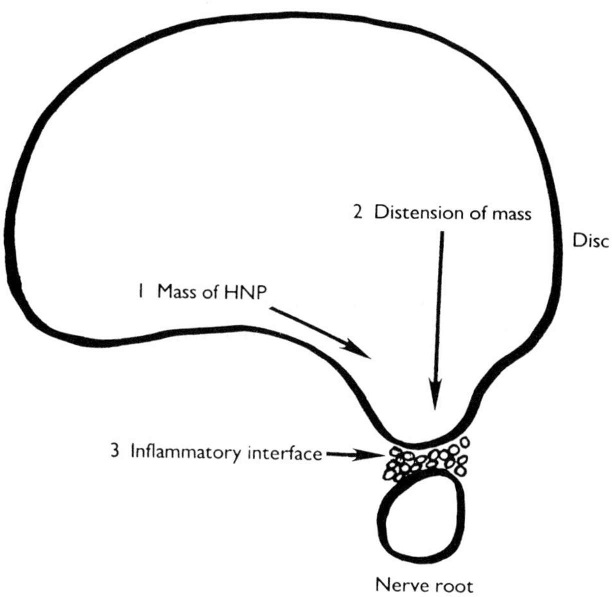

Fig. 8.5
The three local components that contribute to the symptom of sciatica.

of nucleus pulposus can be thought of as collagen-reinforcing bars laid in a viscoelastic base of proteoglycan and water. It is normally a pliable structure but, through the various effects of trauma and ageing, can degenerate or be displaced, in whole or in part, from the disc space cavity into the spinal canal.

Chymopapain's only effect is on the proteoglycan–water aggregates of nuclear tissue[7] (Fig. 8.4). This proteoglycan aggregate imbibes water and cements itself amongst the collagen fibres to give the nucleus pulposus its viscoelastic structure. Because of the chemical make-up of proteoglycan and its fluid nature, it has the ability to absorb and release water, thereby absorbing the dissipating forces across the disc space. Proteoglycans are negatively charged, whilst chymopapain is positively charged.

During displacement of discal nuclear material into the spinal canal, three local components may contribute to the symptom of sciatica (Fig. 8.5). Chymopapain, positively charged, has a direct effect on the negatively charged proteoglycan aggregate by splitting off the glycosaminoglycan side-chains. This in turn interferes with the ability of proteoglycan to hold water. The hydrolysis deflates the nuclear bulge and thus reduces pressure on the nerve root. Since collagen is not directly affected, there is still some mass of disc tissue present for weeks or months after chemonucleolysis[8] (Fig. 8.6). This has been recently demonstrated by Kato and coworkers,[9] who showed persistent MRI defects after successful chemonucleolysis. They felt that after 4 weeks and up until 3 months post-injection, scarring of the intradiscal and extruded material was complete. They erroneously concluded that the success of chemonucleolysis was due to this scarring when in fact other studies[8,10–12] suggest that the immediate deflation of the distended disc relieves pressure on the nerve root, which immediately decreases sciatic symptoms.

To understand the failure of the action of chymopapain in the extruded/sequestered disc herniation, one must appreciate that in addition to the relatively inaccessible location of the disc fragment to chymopapain, the histological make-up of this mass of material lying in the spinal canal is almost uniformly collagenous and thus unaffected by the enzyme. At the other end of the spectrum is a contained disc protrusion, potentially made up of a considerable amount of proteoglycan; it responds dramatically to chymopapain. The combination of a displaced fragment and the collagenous nature of that fragment makes chymopapain injection for an extruded/sequestered disc fruitless.

Toxicology

Except in subarachnoid injection, chymopapain has a wide margin of safety between the effective therapeutic dose and the toxic dose.[13] In animal experiments there is a 100-fold margin of safety between the effective therapeutic dose and the toxic dose. Because chymopapain has no effect on collagenase structures such as bone, ligaments, muscle, nerve and epidural tissue, it is safe and has virtually no local complications when properly injected. Rydevik[14] and associates reported adverse effects on nerve tissue in rabbit tibial nerves, but this has not been borne out by clinical work.[15]

Chymopapain is dangerous when injected into the subarachnoid space. It dissolves the basement membrane

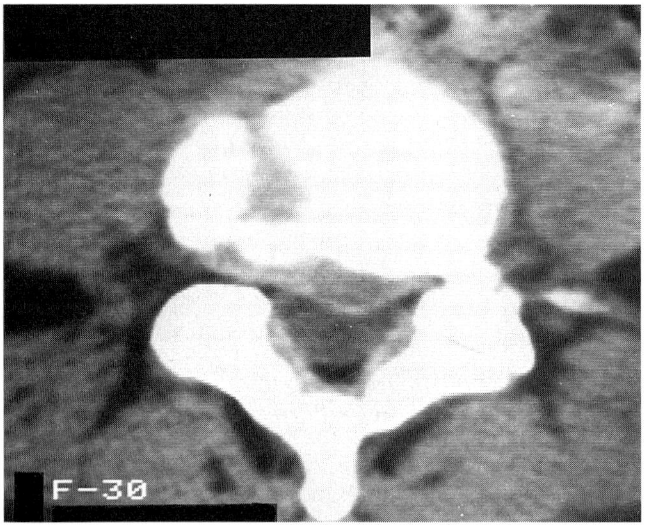

(a)

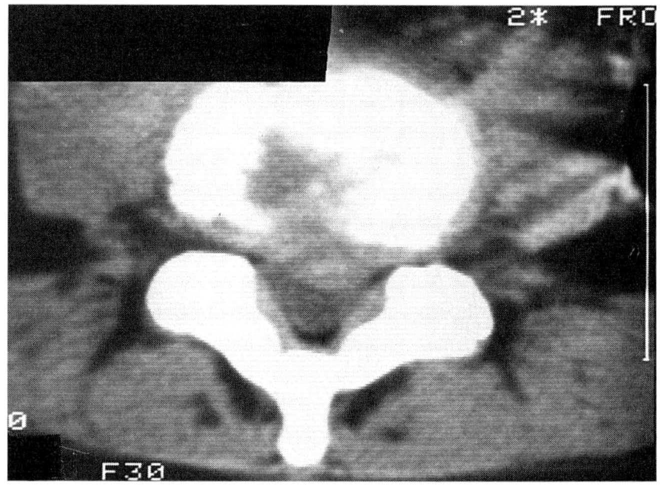

(b)

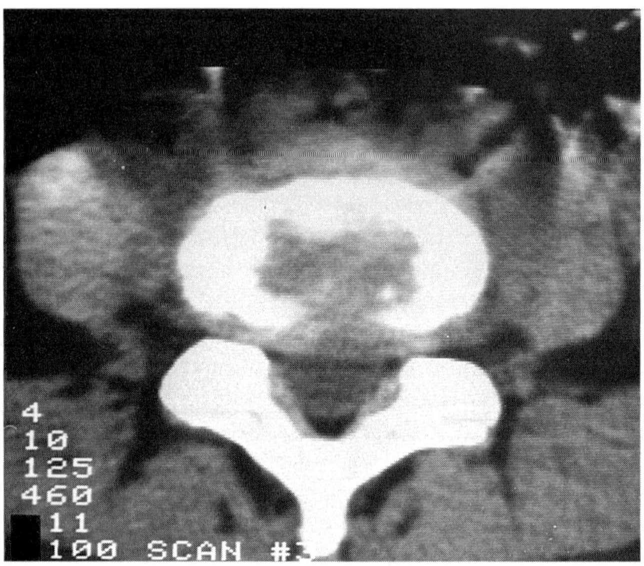

(c)

Fig. 8.6
(a) CT before chemonucleolysis. (b) One month after successful chemonucleolysis. Note the persistence of mass in region of left L5–S1 HNP, yet patient's leg pain has been relieved. (c) Six months later, with no further treatment intervention, the residual collagen mass has disappeared.

of the pia-arachnoid vessels, resulting in subarachnoid haemorrhage. This haemorrhage can cause a local phenomenon such as arachnoiditis and cauda equina syndrome,[16–20] or it can spread through the entire cerebrospinal fluid (CSF) space and cause subarachnoid haemorrhage.

In addition, there seems to be a delayed serious, albeit iatrogenic, possibly hypersensitive effect of chymopapain on the subarachnoid space. This effect has caused transverse myelitis[21] in humans. These topics will be covered in the section on chymopapain complications.

Indications for chemonucleolysis

Table 8.2 lists the indications for interventional therapy in lumbar disc prolapse. Chemonucleolysis is indicated only for patients in the relative group of indications. It is strongly contraindicated in the absolute group of indications for surgery.

Within the group of patients with relative indications, a subgroup of extruded/sequestered discs needs to be excluded. This is a diagnosis that can be made only on investigation and not clinically.[22] The following findings on investigation are suggestive of either an extruded/sequestered disc or a chymopapain non-responsive disc herniation:

1 large HNP (filling more than 50 per cent of the sagittal canal diameter on any one MRI or CT scan slice);
2 a pedunculated shape to the disc fragment (a disc fragment with a long sagittal dimension and a narrow stalk or coronal dimension);
3 a disc fragment that has migrated away from the disc space;
4 artefactual changes, such as calcification (osteophyte formation) or air in the fragment.

The following investigative findings suggest that a disc protrusion is present and chymopapain injection may be of benefit:

1 a defect exactly opposite the disc space;
2 a shallow defect with more coronal breadth than sagittal depth (on axial CT or MRI) (Fig.8.4).

Contraindications to Chemonucleolysis

Table 8.3 summarizes the contraindications, both relative and absolute, to the use of chemonucleosis.

Absolute
Chymopapain allergy (skin, serum tests)
Pregnancy
Cauda equina syndrome
Significant neurological deficit and large fragment of herniated disc on MRI
Relative
Cord levels
Prior chymopapain injection
Extruded/sequestered disc
Previous surgery at same level
Spinal stenosis
A dominant symptom of back pain
Spondylolisthesis
Psychogenic pain component

Table 8.3 *Contraindications to chemonucleolysis*

Relative contraindications

SPINAL STENOSIS

Spinal stenosis affecting the canal or lateral zone never responds to chemonucleolysis. Occasionally these bony defects are complicated by a disc herniation; it is highly unlikely that these combined conditions will be favourably altered by chemonucleolysis. In fact, the post-injection disc space narrowing may aggravate the spinal canal or lateral zone stenosis. This occurrence has led to the false assumption that chymopapain causes spinal stenosis, when in fact stenosis was present but undiagnosed before chemonucleolysis and became a cause of persisting symptoms after injection. The wider use of CT scanning and MRI to screen patients before chemonucleolysis is eliminating this problem.

PREVIOUS SURGERY

Patients who have had previous surgery and suffer from recurrent disc herniations at the same level and same side should not have chemonucleolysis. Patients who have had previous surgery and suffer from a disc herniation at a different level or opposite side may benefit from chemonucleolysis.

Strong contraindications

DEGENERATIVE DISC DISEASE AND FACET JOINT DISEASE

Patients with back pain predominantly caused by mechanical instability, either within the disc space[23] or within the facet joint, do not respond to chymopapain injection.

SPINAL INSTABILITY

Patients with spondylolisthesis do not benefit from the injection of chymopapain at the slip level. In fact, the temporary instability that occurs at a disc space level after the injection of chymopapain could cause an increase in the listhesis. This contraindication applies to both lytic and degenerative spondylolisthesis. A patient with spondylolisthesis may have a symptomatic disc herniation at a level other than the slip level, and the patient would qualify for chymopapain injection, provided they fulfil the criteria for diagnosis of a soft disc herniation causing sciatica.

Absolute contraindications

SENSITIVITY TO CHYMOPAPAIN

A patient may give a history of an allergy to the ingestion of meat tenderizer or foods containing chymopapain (beer, cheese, and some toothpastes). These patients are at risk for an allergic reaction and should not be considered for chymopapain injection. A positive skin test[24] for chymopapain sensitivity is an obvious contraindication to chymopapain injection. A patient who has had a previous chemonucleolysis may, after a long latent interval, develop another disc herniation at another level. Manufacturers caution against a second chymopapain injection.[25] If that patient has a negative skin test, however, a repeat chemonucleolysis (different level) could be considered.[24]

DISC HERNIATIONS AT CORD LEVELS

To date there is no recommendation for chymopapain to be injected into disc herniation at cord levels. The cord level could be either a thoracic or cervical disc. Unfortunately most thoracic discs are calcified, and most cervical radicular syndromes are chronic with an element of bony nerve root entrapment, and a chymopapain injection would be of no benefit. There is ongoing research into the use of chymopapain in the treatment of the infrequent soft cervical disc herniation, and a recommendation for or against injection into soft cervical or thoracic discs cannot be made at present.

NEUROLOGICAL LESION OF UNKNOWN CAUSE

It is unreasonable to risk plant enzyme injection in a patient who has a neurological problem of obscure cause, such as multiple sclerosis.

PREGNANT WOMEN

The effect of chymopapain on the fetus has never been documented, and its use in pregnancy is not recommended.

NON-ORGANIC SPINAL PAIN

Patients with emotional or functional causes of back disability will not benefit from chemonucleolysis.

PATIENT AGE AND DURATION OF SYMPTOMS

Younger patients (below the age of 25) are more likely to have a simple disc protrusion, whereas the more mature patient has a higher incidence of disc extrusion or sequestration. There is a tendency for the older patient or the patient with many years of symptoms to have associated diagnoses such as spinal stenosis. The adult (average age 40) and the older patient (over 55) often have pathology that does not respond to chymopapain; thus, the most common use of chemonucleolysis is in the younger patient with a simple disc protrusion.[26]

TECHNIQUE OF CHEMONUCLEOLYSIS

Considerable confusion arose when one manufacturer[25,27,28] originally recommended the use of general anaesthesia, premedication, a needle insertion site 8 cm from the mid-line, a 45° angle to approach the disc space, single-needle technique to get into the L5–S1 disc space, and the use of contrast material for discographic assessment for disc integrity. After the complications that followed the release of chymopapain in the United States, a critical review suggested that a safer protocol was in order, resulting in recommended changes in the technique. These changes essentially repeat the recommendations of numerous authors,[13,29–32] and include the use of local neuroleptic anaesthesia, the selection of a needle insertion site at an adequate distance from the mid-line (10 cm average), the use of a double-needle technique for at least the L5–S1 disc space, and the avoidance of discography using contrast material.

Whether or not contrast material is used, the basic technical rule is that you must know exactly where the needle tip is before injecting chymopapain. If this requires the use of contrast material and a screening roentgenogram, contrast material must be used. With proper needle insertion technique and two good radiographs at 90° to each other, routine use of contrast material is not necessary to localize needle tip position.

Local versus general anaesthesia

To the author's knowledge, no death has occurred in a patient undergoing chemonucleolysis under local anaesthesia; all deaths following injection of chymopapain have occurred in patients who underwent the procedure under general anaesthesia.[16,33] A review of many of these deaths revealed technical complication facilitated by an unresponsive patient under general anaesthesia. The procedure must be done under local anaesthesia, with or without the use of a neuroleptic.

Prone or lateral position?

The lateral position is the choice of most users. The prone position is acceptable, provided the puncture site is not too lateral. Theoretically, the bowel has not fallen away from the lateral abdominal gutter in the prone position. In the lateral position, the bowel falls away from the lateral gutter, and a 10-cm insertion site is safe.

Principles of the lateral approach

1. An uncompromised image intensification system is mandatory.
2. Maintain proper position of the patient. If the procedure is being performed with the patient in the lateral position, there is a tendency for the awake patient to roll out of the perfect lateral position. The most common mistake made by the beginner trying to get a needle into a disc space is to fail to observe that the patient has rolled into an oblique position.

3 Select the correct disc. It is important to see the front of the sacrum before selecting the disc space for injection. Without clearly identifying the sacrum one can inadvertently put a needle into the L3–L4 disc space, believing it to be at the L4–L5 level. This can be prevented by viewing the front of the sacrum before counting levels. Four or six lumbar vertebrae or other congenital lumbosacral anomalies can confuse the selection of the disc space to be punctured.

4 Select the correct insertion site. It is important to get far enough from the mid-line. In the average patient 10 cm from the mid-line, adjacent to the iliac crest, is the correct insertion site. In a larger patient one may have to go further than 10 cm from the mid-line, and in a smaller patient a needle insertion site closer to the mid-line should be selected.

5 Approach the disc at the proper angle. The angle of approach to the disc space is closer to 60° than 45° degrees; however, it is important to remember that the needle must not penetrate the foramen and enter the spinal canal to puncture the subarachnoid space. It is advisable to leave the stylet out of the needle as an additional precaution. Inadvertent puncture of the subarachnoid space is then heralded by the immediate back-flow of CSF, signalling the necessity to abort the procedure and reschedule in 2-3 weeks.

6 Position the needle tip in the centre of the disc. The centre of the disc space is defined as the middle third of the disc space on lateral roentgenogram and superimposed on the spinous process on antero-posterior (AP) roentgenogram (Fig 8.7).

7 Discography or discometry. The use of contrast or the water acceptance test will support the fact that the needle tip is within the discal/nuclear cavity.

If more than three passes of the needle have occurred without entry to the disc space being achieved, abort the procedure. Similarly, if you penetrate the nerve root (i.e. cause severe leg pain), abort the procedure. I recommend that after puncturing the skin, the stylet should be removed from the needle and the needle advanced toward the disc space with an open lumen. If you inadvertently cross the subarachnoid space on advancing the needle, the immediate back-flow of CSF will be ample notification to abort the procedure. Never abandon the lateral approach for the mid-line transdural approach.

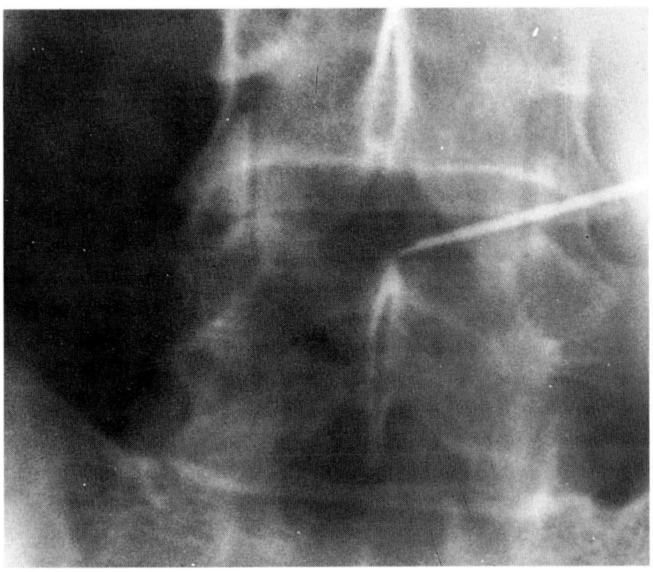

(a)

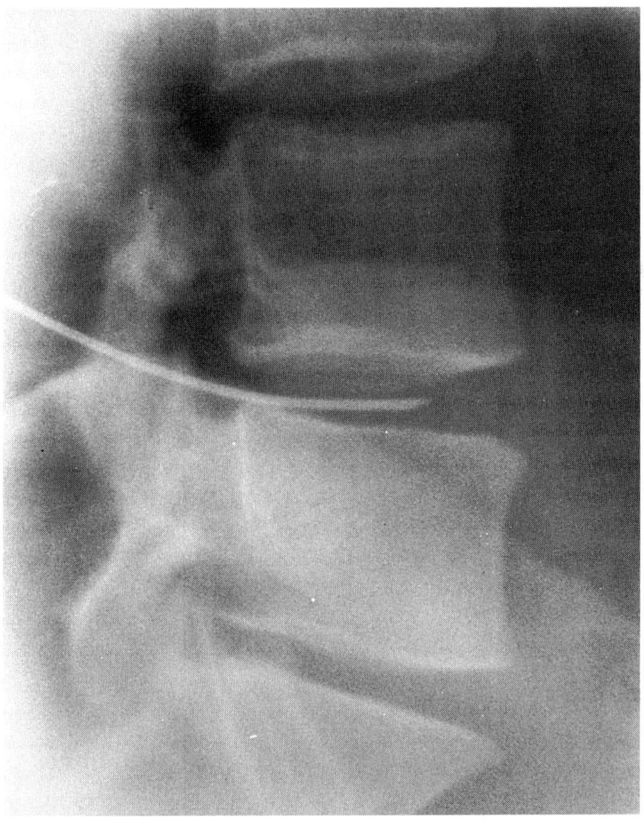

(b)

Fig. 8.7
(a) AP radiograph with good positioning of needle superimposed on spinous process. (b) Lateral radiograph with needle just at the anterior edge of the middle third of the disc – a just barely acceptable position.

8 The proper amount of active chymopapain to be injected into a clean disc space ranges from 0.75 to 1.0 ml (2000 pKat/ml). Benoist and colleagues,[34] in a recent article, have found no difference between low-dose chymopapain (1 ml) and standard dose (2 ml).

Post-operative care

Most patients will notice increased back pain after the injection,[35] and for some it will be severe muscle spasm. Within a few days the pain will be tolerable and controlled by mild analgesia. For more severe muscle spasm pain, intravenous steroid in the immediate post-operative phase or steroid by mouth in the intermediate stage will usually break the cycle of spasm.

Most patients who are heading to a successful result will notice immediate relief of leg pain. Some patients with lingering leg pain will still improve up to 4–6 weeks following the injection and it is reasonable to delay operative intervention until that time. Persisting disabling leg pain at 6 weeks should be considered a failure, with no cause for delaying the decision on surgical intervention.

Follow-up is usually beneficial at 4–6 weeks. Relief of leg pain, good to normal straight leg raising ability and disc space narrowing on plain lateral radiographs are indicators of a good result.

RESULTS

The decisive study supporting the efficacy of chemonucleolysis was completed by Fraser.[36–38] Using carefully selected patients he demonstrated a 75 per cent success rate up to 10 years after chemonucleolysis, compared with a placebo response rate of 50 per cent success in saline-injected controls. Other blinded studies have supported Fraser's conclusions.[29,39,40]

Other publications have reported results varying from poor[41] to 90–95 per cent success.[33] In multiple studies[26,30,42,43] the authors report approximately a 70 per cent good-result rate, slightly higher for younger patients, and slightly better at the L5–S1 level in all age groups.

The most common cause of failure is the sequestered or extruded fragment of disc material that has not been dissolved by chymopapain. Since the nuclear content of the disc space has been dissolved by the chymopapain, little work needs to be done within the disc space itself during surgery for failed chemonucleolysis. In fact, microsurgery[44] dealing with the spinal canal pathology alone is the treatment of choice in these patients. Chemonucleolysis did not compromise surgical intervention in over 300 personal cases.

It is obvious that the 70 per cent good-result rate will be improved on only by correctly identifying patients who have an extruded or sequestered disc and using surgical excision in these cases.

COMPLICATIONS

Anaphylaxis

Anaphylaxis occurs in 0.35 per cent of patients[5] who have not been pre-screened. It is manifested by a profound drop in blood pressure and requires vigorous immediate resuscitation to save the patient's life. The cornerstones of treatment for anaphylaxis are appropriate (large) doses of adrenaline, large volumes of intravenous fluids, steroids and antihistamines.

With the use of skin testing,[24] anaphylaxis is becoming a less frequent problem. The author has completed 950 cases of chymopapain injection preceded by skin testing for chymopapain sensitivity. To date, there has not been an anaphylactic reaction in a patient who has had a negative skin test. Thus the management of anaphylaxis has moved into prophylaxis. Any patient with a positive skin test is eliminated from treatment with chymopapain.

Neurological complications

Chymopapain, with or without contrast material, injected into the subarachnoid space is almost certain to be disastrous.[16, 17] It is extremely important that the needle tip does not cross the subarachnoid space on insertion into the nuclear cavity or reside in the subarachnoid space at the time of injection. If there is any question about the needle tip position, the procedure should be aborted.

The possibility of a connection between the disc space and the subarachnoid space is the one drawback to not using contrast material at the time of needle placement. It is obvious after injecting contrast material into the disc that when it appears in the subarachnoid space, there is a connection between the disc cavity and the subarachnoid space. The most common cause of this connection is crossing the subarachnoid space on needle insertion. It is my opinion that whenever there is any degree of resistance to injection of test material into the nuclear cavity, it is highly unlikely that there is a connection between the disc space and the subarachnoid space. For this reason, it is possible to use a water acceptance test rather than contrast material at the time of chymopapain injection.

Injection of chymopapain (with or without contrast material) into the subarachnoid space may cause subarachnoid haemorrhage with cerebral complications, a cauda equina syndrome or a delayed traverse myelitis. Preliminary experimental work[28] suggests that it is not chymopapain alone but the combination of chymopapain and contrast material in the subarachnoid space that causes serious neurologic complications. Other research suggests that chymopapain alone is capable of causing subarachnoid haemorrhage and serious neurological complications.[18]

Mortality

The author has not had a death in over 8000 injections. Recent publications on surgical discectomy describe a 0.15–1.0 per cent mortality rate;[45–50] Bouillet,[18] in a large European study on chymopapain use, reported a mortality rate of zero in a multicentred survey of 43 622 patient injections.

Miscellaneous complications

Two miscellaneous complications of chemonucleolysis require consideration. The first is root damage from penetration of the nerve root at the time of needle insertion.[32] This usually occurs under general anaesthesia and affects the fifth lumbar nerve root, with damage caused by attempting to get the needle into the L5–S1 disc space. Root penetration has led to some permanent causalgic syndromes. Under local neuroleptic anaesthesia, this is a rare complication.

A second complication is discitis, either low-grade infective discitis or chemical discitis that results in a prolongation of the patient's back pain and a typical roentgenographic picture. Fraser and associates[51,52] have done the definitive work in this area as well. They have shown that although chymopapain is bactericidal *in vitro*, it does not appear to prevent a particular type of disc inflammation *in vivo*. Histological sections of sheep disc spaces injected with a combination of chymopapain and bacteria (*Staphylococcus epidermidis*) showed some signs of disc space infection yet no disc yielded a positive culture. They concluded that most cases of discitis, although low grade in severity and often sterile on culture, originate from bacterial contamination of the disc at the time of the procedure.

Competing procedures

Numerous studies[53–56] have recently appeared comparing chemonucleolysis to open surgery and to percutaneous discectomy. Two things are becoming apparent. Surgery, as expected, delivers consistently better results than chemonucleolysis. If one sees chemonucleolysis as the last step in conservative care, saving 70 per cent of properly selected patients from surgery, chemonucleolysis has found its place. Its role is not that of a competitive procedure to surgery.

The second obvious conclusion is the high failure rate and recurrence rate following percutaneous discectomy, also predictable. With the appearance of these studies there is a renewed interest in chymopapain in North America.

Conclusion

Chemonucleolysis, for the treatment of lumbar disc rupture, is no different from any other medical or surgical procedure in that proper patient selection and careful technique are the key to a high level of success.

References

1. Mixter WJ, Barr JS (1934). Rupture of the intervertebral disc with involvement of the spinal canal. *New England Journal of Medicine* **211**: 210–15.
2. Verbiest H (1954). Radicular syndrome from developmental narrowing of lumbar vertebral canal. *Journal of Bone and Joint Surgery* **36B**: 230–7.
3. Boden SD, Daris DO, Dina TS, Patrohas NJ, Wiesel SW (1990). Abnormal magnetic-resonance scans of the lumbar spine in asymptomatic subjects: A prospective investigation. *Journal of Bone and Joint Surgery* **73**: 403–8.
4. Smith L (1964). Enzyme dissolution of nucleus pulposus in humans. *Journal of the American Medical Association* **187**: 137–40.
5. Hall BB, McCulloch JA (1983). Anaphylactic reactions following the intradiscal injection of chymopapain under local anesthesia. *Journal of Bone and Joint Surgery* **65**(9): 1215–19.
6. Jansen EF, Balls AK (1941). Chymopapain: New crystalline proteinase from papaya latex. *Journal of Biological Chemistry* **137**: 459–63.
7. Wakano K, Kasman R, Chao EY, Bradford DS, Oegema TR (1983). Biochemical analysis of canine intervertebral disc after chymopapain injection: a preliminary report. *Spine* **8**(1): 59–68.
8. McCulloch JA (1984). Computed tomography before and after chemonucleolysis. In: *Computed Tomography of the Spine*. Edited by Post MJD. Williams & Wilkins, Baltimore.
9. Kato F, Mimatsu K, Kawakami N, Iwata H, Miura T (1992). Serial changes observed by magnetic resonance imaging in the intervertebral disc after chemonucleolysis. *Spine* **17**(8): 934–9.

10 Gentry L, Turski PA, Strother CM *et al.* (1985): Chymopapain chemonucleolysis: CT changes after treatment. *AJR* **145**(2): 361–9.
11 Huckman MS, Clark JW, McNeill TW, Whisler WW, Hejna WF, Russell EJ, Ramsey RG, Turner D (1987). Chemonucleation and changes observed on lumbar MR scan: preliminary report. *AJNR* **8**: 1–4.
12 Masaryk TJ, Boumphrey F, Modic MT, Tamborrello C, Ross JS, Brown MD (1986). Effects of chemonucleolysis demonstrated by MR imaging. *Journal of Computer Assisted Tomography* **10**: 917–923.
13 Weiner DS, Macnab I (1970). The use of chymopapain in degenerative disc disease a preliminary report. *Canadian Medical Association Journal* **102**(12): 1252–6.
14 Rydevik B, Branemark P-I, Norberg C, McLean WG, Sjostrand J, Fogelberg M (1976). Effects of chymopapain on nerve tissue. *Spine* **1**: 137–48.
15 Russell AS (1975). Chymopapain chemonucleolysis in lumbar disc disease. *Journal of the American Medical Association* **233**(11): 1164.
16 Agre K, Wilson RR, Brim M, McDermott DJ (1984). Chymodiactin post-marketing surveillance. *Spine* **9**: 479.
17 Bouillet R (1983). Complications of discal hernia therapy: Comparative study regarding surgical therapy and nucleolysis by chymopapain. *Acta Orthopaedica Belgica* **49** (Suppl.): 48–77
18 Bouillet R (1990). Treatment of sciatica: A comparative survey of complications of surgical treatment and nucleolysis with chymopapain. *Clinical Orthopaedics and Related Research* **251**: 144–52.
19 Kitchel SH, Brown MD (1992). Complications of chemonucleolysis. *Clinical Orthopaedics and Related Research* **284**: 63–74.
20 Morris J (1984). Complications of chemonucleolysis. Presented at Spine Update, San Francisco.
21 Eguro H (1983). Transverse myelitis following chemonucleolysis: Report of a case. *Journal of Bone and Joint Surgery* **65A**: 1328–30.
22 McCulloch JA, Lambe D: Unpublished data.
23 Katz ME, Teitelbaum SL, Gilula LA, Resnick D, Katz SJ (1988). Radiologic and pathologic patterns of end-plate-based vertebral sclerosis. *Investigative Radiology* **23**(6): 447–54.
24 McCulloch JA, Dolovich G, Canham W (1985). Skin testing for chymopapain allergy: A preliminary report. *Annals of Allergy* **55**: 609–11.
25 Smith Laboratories, Inc., Northbrook, IL: Original product brochure.
26 Lorenz M, McCulloch JA (1985). Chemonucleolysis for herniated nucleus pulposus in adolescents. *Journal of Bone and Joint Surgery* **67A**: 1402–4.
27 Smith Laboratories, Inc., Northbrook, IL: Advertising brochure.
28 Smith Laboratories, Inc., Northbrook, IL: Product information letter, July 1984.
29 Dabezies EJ, Langford K, Morris J, Sields CB, Wilkinson HA (1988). Safety and efficacy of chymopapain (discase) in the treatment of sciatica due to a herniated nucleus pulposus: Results of a randomized, double-blinded study. *Spine* **13**(5): 561–5.
30 McCulloch JA (1981). Chemonucleolysis for relief of sciatica due to a herniated intervertebral disc. *Canadian Medical Association Journal* **124**(7): 879–82.
31 Parkinson D, Shields C (1973). Treatment of protruded lumbar intervertebral discs with chymopapain. *Journal of Neurosurgery* **39**(2): 203–8.
32 Wiltse LL, Widell EH Jr, Yuan HA (1975). Chymopapain chemonucleolysis in lumbar disc disease. *Journal of the American Medical Association* **231**(5):474–9.
33 Smith Laboratories, Inc., Northbrook, IL (1985). Data from postmarketing surveillance.
34 Benoist M, Bonneville J, Lassale B. *et al.* (1993). A randomized, double-blind study to compare low-dose with standard dose chymopapain in the treatment of herniated lumbar intervertebral discs. *Spine* **18**: 28–34.
35 Brown MD, Tompkins JS (1989). Pain response post-chemonucleolysis or disc excision. *Spine* **14**: 321–6.
36 Fraser RD (1982). Chymopapain for the treatment of intervertebral disc herniation: A preliminary report of a double-blinded study. *Spine* **7**(6): 608–12.
37 Fraser RD (1984). Chymopapain for the treatment of intervertebral disc herniation: The final report of the double-blinded study. *Spine* **9**(8): 815–18.
38 Gogan WJ, Fraser RD (1992). Chymopapain: A 10-year double-blind study. *Spine* **17**: 388–94.
39 Javid MJ, Nordby EJ, Ford LT, Hejna WJ, Whisler WW, Burton C, Millett DK, Wiltse LL, Widell EM Jr, Boyd RJ, Newton SE, Thisted R (1983). Safety and efficacy of chymopapain (chymodiactin) in herniated nucleus pulposus with sciatica. *Journal of the American Medical Association* **249**: 2489–94.
40 Travenol Laboratories, Inc,: New Drug Application 18–625. Submitted to United States Food and Drug Administration, April 1981.
41 Schwetschenau PR, Ramirez A, Johnston J, Wiggs C, Martins AN (1976). Double-blind evaluations of intradiscal chymopapain for herniated lumbar discs: Early results. *Journal of Neurosurgery* **45**: 622–7.
42 McCulloch JA (1983). Outpatient discolysis with chymopapain. *Orthopedics* **6**: 1624–7.
43 Tregonning GD, Transfeldt EE, McCulloch JA, Macnab I, Nachemson A (1991). Chymopapain versus conventional surgery for lumbar disc herniation. *Journal of Bone and Joint Surgery* **73B**: 481–6.
44 Kitaoka H, McCulloch JA (1988). Microdiscectomy for failed chemonucleolysis. *Neuro-Orthopedics* **5**: 45–51.
45 Mayfield FH (1976). Complications of laminectomy. *Clinical Neurosurgery* **23**: 435–9.
46 McLaren AC, Bailey SI (1986). Cauda equina syndrome: A complication of lumbar discectomy. *Clinical Orthopaedics and Related Research* **204**: 143–9.
47 Pappas CTE, Harrington T, Sonntag VKH (1992). Outcome analysis of 654 surgically treated lumbar disc herniations. *Neurosurgery* **30**(6): 862–6.
48 Ramirez LF, Thisled R (1989). Using a National Health Care data base to determine surgical complications in community hospitals: Lumbar discectomy as an example. *Neurosurgery* **25**: 218–25.
49 Rish BL (1984). A critique of the surgical management of

lumbar disc disease in a private neurosurgical practice. *Spine* **9**(5): 500–4.

50 SPANGFORT EV (1972). The lumbar disc herniation: a computer-aided analysis of 2504 operations. *Acta Orthopaedica Scandinavica* **142** (Suppl): 1–95.

51 FRASER RD, OSTI O, VERNON-ROBERTS B (1987). Discitis after discography 1: Incidence and pathological findings. *Journal of Bone and Joint Surgery* **69B**: 26–35.

52 FRASER RD, OSTI OL, VERNON-ROBERTS B (1986). 1986 Volvo Award in basic science: discitis following chemonucleolysis: an experimental study. *Spine* **11**(7): 679–87.

53 JAVID MJ (1992). A 1 to 4 year follow-up review of treatment of sciatica using chemonucleolysis or laminectomy. *Journal of Neurosurgery* **76**: 184–90.

54 MURALIKUTTAN KP, HAMILTON A, KERNOHAN WG, MOLLAN RA, ADAIR IV (1992). A prospective randomized trial of chemonucleolysis and conventional disc surgery in single level lumbar disc herniation. *Spine* **17**(4): 381–7.

55 REVEL M, PAYAN C, VALLEE C, LAREDO JD, LASSALE B, ROUX C, CARTER H, SALOMON C, DELMAS E, ROUCOULES J, *et al.* (1993). Automated percutaneous lumbar discectomy versus chemonucleolysis in the treatment of sciatica: A randomized multicenter trial. *Spine* **18**: 1:1–7.

56 WEINSTEIN J, SPRATT KF, LEHMANN T, MCNEILL T, HEJNA W (1986). Lumbar disc herniation: A comparison of the results of chemonucleolysis and open discectomy after 10 years. *Journal of Bone and Joint Surgery* **68**(1): 43–54.

CHAPTER 9

Nucleotomy

Gary Onik

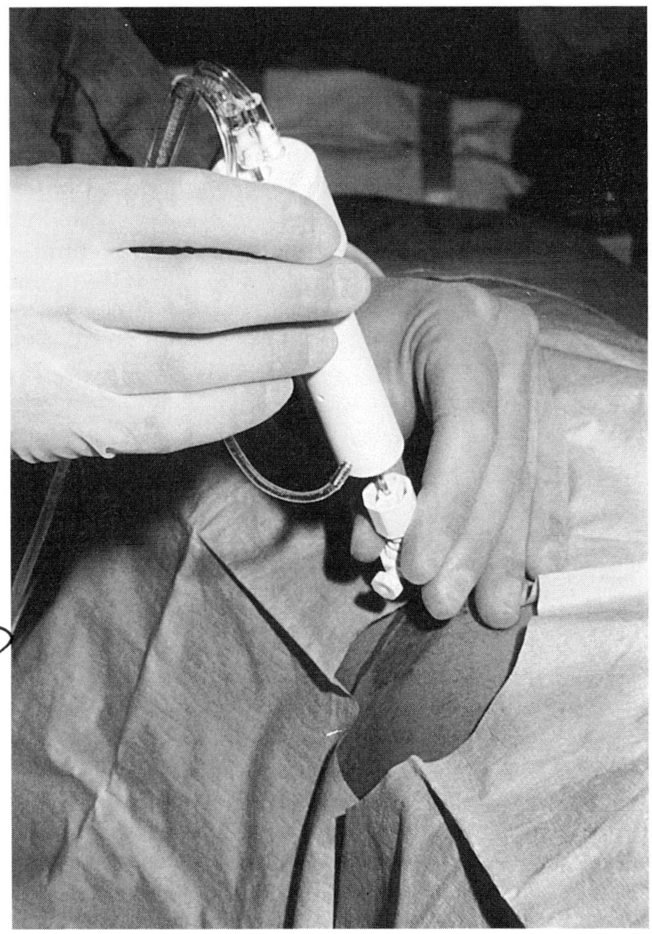

Introduction

The growth of percutaneous intradiscal therapy as an alternative to the traditional open laminectomy and microdiscectomy was a result of the perception of both the public and many in the medical profession that the risks of open disc surgery were great and that an alternative was needed. The principle of percutaneous intradiscal therapy is to decrease disruption of the posterior and extradural elements of the spine, thereby decreasing morbidity. One of the first intradiscal therapies was manual percutaneous discectomy, which was an attempt to provide a treatment as successful as open discectomy and chemonucleolysis, but with a lower morbidity.

Hijikata *et al.* in 1975 were the first to report the manual method of percutaneous discectomy. A large tube or cannula (up to 5 mm) was placed down to the annulus and the disc material removed by the use of rongeurs in the traditional surgical manner[1]. In 1989 Hijikata reported his 14-year experience with a success rate comparable to that of chemonucleolysis.[2] This study, however, still reported major complications including discitis and a major vascular injury. Hijikata's technique has been modified by a number of investigators, but none of these modifications eliminated these complications.[3, 4]

Automated percutaneous lumbar discectomy (APLD) was developed to improve upon the inadequacies of the manual method. Using an automated instrument to remove the disc material provides rapid removal of nucleus pulposus, while giving up to a fourfold decrease in a cross-sectional area (Fig. 9.1). By decreasing the size of the instruments and using a blunt-tipped instrument unable to cut annulus or push its way out of the disc space, the procedure has less chance of major complications. With the instrument requiring only a single placement into the disc, the infection rate is decreased. These equipment changes, coupled with a standardized procedure and reliable radiographic landmarks, resulted in a procedure with minimal morbidity.

It is clear from the literature that APLD is the safest procedure that can be performed on patients for a herniated nucleus pulposus. These conclusions can be drawn from over 50 separate articles published, comprising well over 5000 patients.[5–20] From this literature, it can be concluded that APLD has fulfilled the most important criterion of a non-invasive procedure, that of safety. In all these published articles, no major complications were reported. A discitis rate of approximately 0.2 per cent, three to five times lower than that of open discectomy, was the only significant inherent morbidity. In addition to a low rate of complication, the results of the procedure have consistently been reported to be between 75 and 85 per cent. Remarkably, there is yet to be a reported death from APLD. It is this safety record that makes APLD the procedure of first choice in patients with contained herniations.[10,21,22]

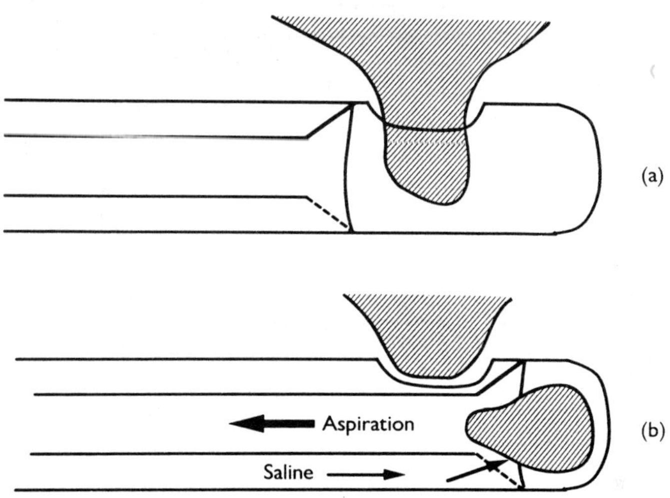

Fig. 9.1
Action of the Nucleotome. Suction is applied to the central bore, thereby aspirating disc material into the side port. Then an inner cutting sleeve is pneumatically driven across the side port, effectively guillotining the disc material. Finally, disc material is suspended in saline, which flows between the walls of the Nucleotome and inner cutting sleeve, and is aspirated away.

Other treatments

In the past, the only acceptable treatment for herniated lumbar discs was open discectomy. However, open discectomy has been associated with high morbidity. Reported intraoperative complications included perforation of the dura, significant haemorrhage, nerve root damage, exposure of the wrong interspace, discitis, postoperative pseudo-meningoceles, arterial injuries and small-bowel injuries.

Chymopapain treatment was the first attempt at intradiscal therapy to decrease the morbidity associated with the operative treatment of herniated discs. After an initial large patient experience, the use of chymopapin has markedly decreased owing to the related complications of the procedure, which included anaphylaxis, neurological complications such as cerebral haemorrage, seizures, paraplegia and quadraplegia, and a significant incidence (from 30 per cent to 40 per cent) of severe postoperative back spasm.[23,24]

Laser discectomy is the newest entry into the intradiscal therapy arena. It involves the vaporization of disc leading to a decrease in volume and therefore a decrease in intradiscal pressure. The major drawback of laser discectomy appears to be the inherent lack of safety of the procedure owing to the spread of the heat beyond the confines of the placement of the instrument.[25] There are reports of injuries to the nerve roots and the bowel.[26,27]

PATIENT SELECTION

The results obtained from percutaneous discectomy vary with the selection of the patients. A large body of literature emphasizes that the procedure is used in the patient with the classic radiograph and clinical findings associated with a focal herniated nucleus pulposus. These patients should have sciatica with leg pain greater than back pain and the classic neurologic finding of wasting, weakness, sensory, reflex changes, and a positive straight leg raising. They must also have failed an adequate course of conservative care which includes physical therapy, oral anti-inflammatories, and possibly epidural steroids.

In addition, these patients should demonstrate a focal herniated nucleus pulposus on MRI or CT that is consistent with their clinical picture (Fig. 9.2). The criteria of a contained herniation include size (less than 50 per cent of the thecal sac), shape of herniation (smooth obtuse margins), intact annular fibres on MRI and migration (≤3 mm).

The most accurate means of determining the contained herniation, however, is the CT discogram (Fig. 9.3). Free flow of contrast into the epidural space indicates a complete tear of the annulus and posterior longitudinal ligament complex. Such patients should be excluded from having the procedure owing to the high likelihood of an extruded fragment (Fig. 9.4).

A second patient group in which the contrast leaks behind the posterior longitudinal ligament, but the disc can be pressurized, has been shown to have a success rate of approximately 50 per cent. With the demonstrated low morbidity of APLD, we will now offer these patients APLD after informing them of these findings and success rate. Even with this success rate, no patient has refused APLD.

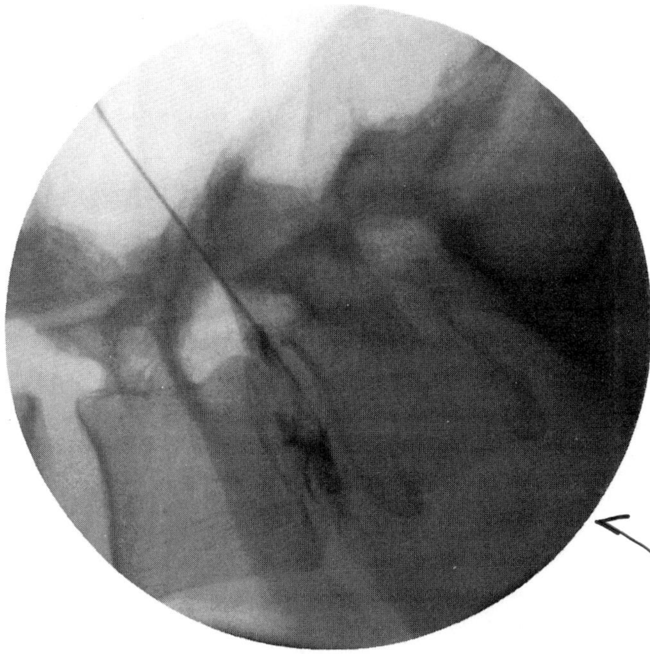

Fig. 9.3
CT discogram of contained HNP. All contrast medium is contained by annulus and posterior longitudinal ligament. No contrast is seen in the epidural space.

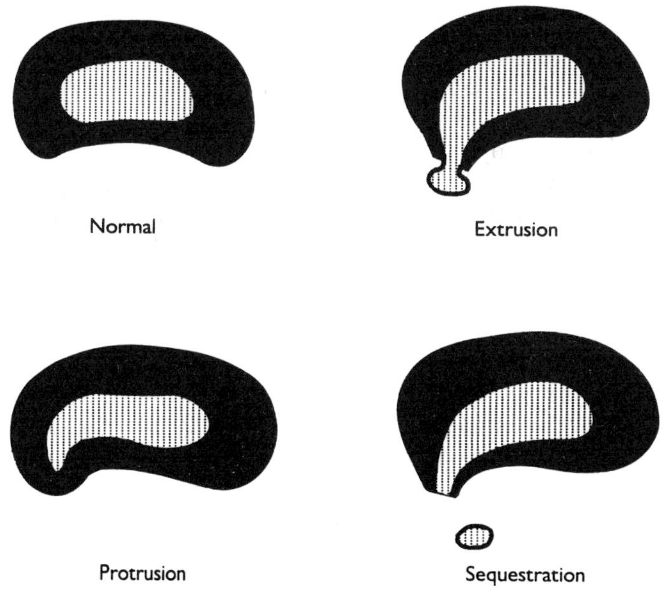

Fig. 9.2
Varying degrees of nuclear herniation. Extruded discs and free fragments are not good candidates for a percutaneous discectomy.

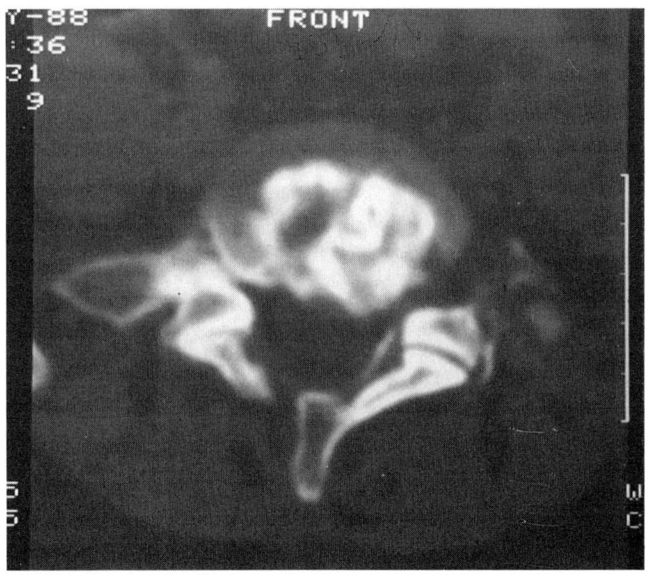

Fig. 9.4
Annulus posterior longitudinal ligament with a complete tear: contrast is flowing out of the disc through the rent in the annulus freely into the epidural space. The disc cannot be pressurized.

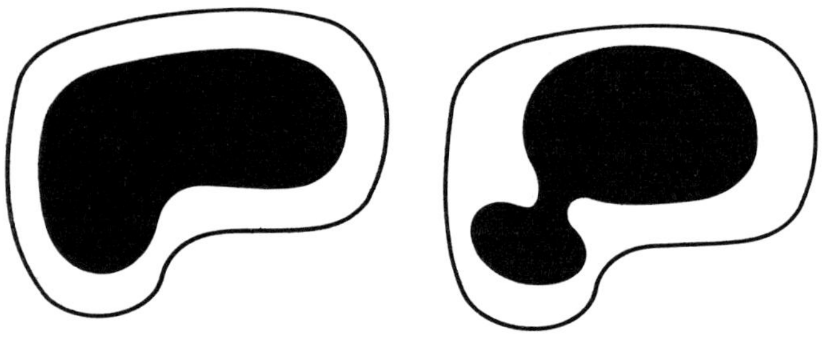

Fig. 9.5
Discographic appearance of annular tears. The right-hand diagram shows a herniation with a narrow tear in the annulus which is essentially a mushroom effect caused when the nucleus herniates through the annulus and is still contained on the opposite side of the rent by the anulus and/or posterior longitudinal ligament. For this herniation treatment with APLD may be inappropriate. The left-hand diagram shows a wide-mouthed herniation that would easily allow retraction of the herniation into the centre of the disc and transmission of pressure differences from the centre of the disc through the rent into the hernition. This is an excellent patient for APLD.

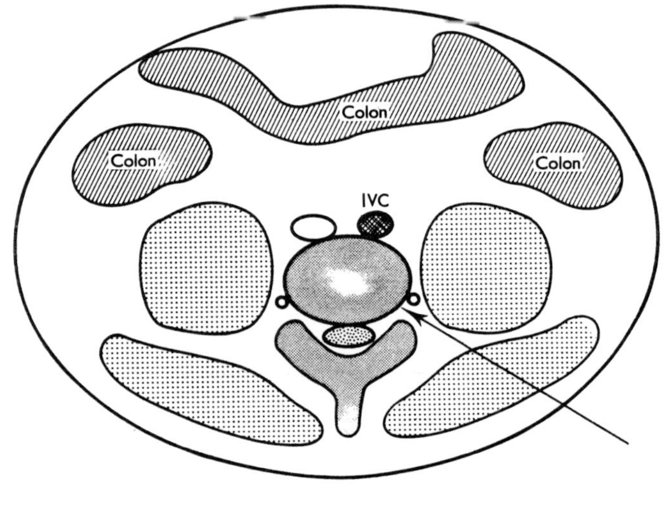

Fig. 9.6
The importance of the posterior vertebral body line. From a postero-lateral approach when the trocar is against the annulus and in the correct trajectory for the centre of the disc, the tip of the trocar will be at a line that connects the posterior vertebral bodies. It is important to note that the structures that one does not want to hit (nerve root, bowel and great vessels) are anterior to this line; therefore, on fluoroscopy when monitoring the procedure, the instrument should never be brought anterior to the posterior vertebral body line unless the annulus is encountered first.

On CT discography, another criterion that should be considered is the size of the rent in the annulus that communicates with the herniation (Fig. 9.5). A wide annular tear makes the success of the procedure likely, while a narrow communication giving a mushroom effect makes transmission of a decreased pressure between the centre of the disc and the herniation unlikely and has been shown to correlate with poor results.[28]

Methods/Technique

The procedure is carried out under under local anaesthesia, since there is a risk of neural injury with general anaesthesia. Using sterile technique the patient is placed in either the lateral dicubitus or the prone position. The entry point is chosen from a preoperative localization CT scan. By means of this postero-lateral approach, an 18 G needle is placed down to the disc space. The placement of the trocar (lateral view) should be parallel to and midway between the vertebral body endplates, with the tip of the trocar directed toward the disc centre. The point should be touching the posterior vertebral body line when the annulus is felt. If the trocar is anterior to this line and the annulus is felt or if the patient is experiencing any radicular pain, the trocar is withdrawn and redirected. When the annulus is felt and the tip of the trocar is at the posterior vertebral body line, the AP view is checked to confirm that the tip is not medial to a line connecting the medial aspect of the pedicles, and is therefore not traversing the thecal sac (Fig. 9.6).

When the trocar is properly placed, the tip is advanced into the disc centre on both the AP and lateral views. The cannula and dilator are inserted over the trocar. They are then pushed down to the wall of the annulus, and their position is confirmed radiographically in two views. Once placement is confirmed, the dilator is removed leaving the trocar and cannula in place. The trephine is placed over the trocar and through the cannula. At this point, the fluoroscopic unit is perpendicular to the cannula and trephine, confirming that they are actually against the annulus.

After the disc is incised, the trephine and trocar are removed leaving the cannula in place and the Nucleotome aspiration probe is placed into the disc. The Nucleotome position is confirmed and it is turned on to aspirate the disc material. After no more disc material can be removed, the rigid cannula can then be angled to change the position of the Nucleotome within the disc. When no further disc material can be removed, the Nucleotome is turned off and pulled back into the cannula and both are removed. The puncture site is then covered with a bandage. In most cases, the patient is held for 3 hours and then discharged the same day of surgery.

Results

Success for patients who have undergone an APLD was defined as the decrease in or resolution of pain and the lack of need for further treatment at the level of the disc herniation. There is now extensive literature regarding APLD from which conclusions about the efficacy of the procedure can be drawn. There is little doubt now that APLD, when performed on the correct patient population, is efficacious, with success rates reported consistently in the 70 per cent to 85 per cent range.[5,12,29–31]

The first major article published on APLD was a prospective multi-institutional study by Onik and Helms which covered over 500 patients.[32] The inclusion criterion was that of the classic 'disc herniation' patient. When the patients met this criterion, a 75 per cent success rate was reported, with failures being due to unrecognized free fragments. Only 15 per cent of patients required an open discectomy. The inclusion of patients who were successfully treated by a second percutaneous discectomy increased the success rate to 80 per cent. These findings were confirmed by a similar multi-institutional study carried out by Bocchi *et al.*, in which a 72 per cent success rate was obtained in over 600 patients.[33] Again, the only complication was one case of discitis. Long-term results after APLD are now being published. In these studies, patients had been followed between 2 and 4.5 years. Success rates ranged from 70 to 78 per cent with no complications[12,34,35]

A number of studies have been published comparing APLD with various other treatment methods. APLD has been compared to chymopapain or chemonucleolysis in at least four different series. In three of these series success rates for APLD and chymopapain were 82 per cent and 72–90 per cent, respectively.[36–38] Because of the lack of back spasms, APLD patients had a decreased hospital stay and an earlier return to function/work.

The only negative report comparing chemonucleolysis and APLD was an article published by Revel *et al.* in which APLD had a 40 per cent success rate compared to 60 per cent for chemonucleolysis.[39] A low success rate for APLD in this study is not surprising since none of the selection criteria were followed in this study including that of excluding migrated fragments of disc. This study basically showed that in a totally unscreened population, 40 per cent of patients could still be successfully treated with APLD.

There are now two studies which compare open surgery and APLD.[21,40] In these reports, APLD had success rates of 72 per cent and 80 per cent as compared to an 80 per cent success rate for an open procedure. APLD, however, resulted in a shorter hospital stay, quicker return to work, and less need for postoperative physical therapy, and because of these findings was considered to be the procedure of first choice.

In conclusion, APLD has more clinical experience than other intradiscal therapy and has demonstrated a superior safety record, making it the procedure of choice. One hopes that continued research and development in this field will expand the patient population and decrease the overall morbidity associated with the present open surgical treatment of herniated lumbar discs.

References

1 Hijikata S, Yamagishi M, Nakayama T *et al.* (1975). Percutaneous discectomy: A new treatment method for lumbar disc herniation. *Journal of Toden Hospital* **5**: 5–13.

2 Hijikata S (1989). Percutaneous nucleotomy: A new concept technique and 12 years' experience. *Clinical Orthopaedics* **238**: 9–23.

3 Schreiber A, Suezawa Y (1986). Transdiscoscopic percutaneous nucleotomy in disk herniation. *Orthopaedic Review* **15**(1): 75–8.

4 Schaffer JL, Kambin P (1991). Percutaneous posterolateral lumbar discectomy and decompression with a 6.9-millimeter cannula. *Journal of Bone and Joint Surgery (American volume)* **73A**(6): 822–31.

5 Onik GM, Mooney Y, Maroon JC *et al.* (1990). Automated percutaneous discectomy: A prospective multi-institutional study. *Neurosurgery* **26**(2): 228–33.

6 Maroon JC, Allen RC (1989). A retrospective study of 1054 APLD cases: A twenty-month clinical follow-up at 35 US centers. *Journal of Neurological and Orthopaedic Medicine and Surgery* **10**(4): L335–7.

7 Benazet JP, Roy-Camille R, Saillant G *et al.* (1991). Treatment of complete lumbar disk herniation by percutaneous discectomy. *Chirurgie (Paris)* **117**(1): 59–67.

8 Blanc C, Meyer A, Tang YS *et al.* (1990). Treatment of herniated lumbar disc by percutaneous nucleotomy with aspiration: Preliminary results in 70 cases. *Journal of Neuroradiology* **17**(3): 182–9.

9 Bocchi L, Ferrata P, Passarello F (1991). The Onik method of automated percutaneous lumbar diskectomy (APLD): Criteria of selection, technique, and evaluation of results. *Italian Journal of Orthopaedics and Traumatology* **17**(1): 5–21.

10 Bonaldi G, Belloni G, Prosetti D *et al.* (1991). Percutaneous discectomy using Onik's method: Three years' experience. *Neuroradiology* **33**: 516–19.

11 Chen SC, Chatterjl S, Solanki R (1992). Automated percutaneous lumbar discectomy: Experience in a district hospital. *Orthopaedic Products News* 28–30.

12 Gill K, Blumenthal SL (1991). Clinical experience with automated percutaneous diskectomy – the Nucleotome system. *Orthopedics* **14**(7): 757–60.

13 Goldstein TB, Mink JH, Dawson EG (1989). Early experience with automated percutaneous lumbar diskectomy in the treatment of lumbar disk herniations. *Clinical Orthopaedics* **238**: 77–82.

14 Hidalgo Ovejero AM, Garcia Mata S, Antunano Zarraga P *et al.* (1991). Resultados iniciales de la nucleotomia percutánea automatizada en el tratemiento de las hernias discales. *Reviews in Orthopaedic Traumatology* **35**(6): 428–33.

15 Kaps H, Cotta H (1989). Early results of automated percutaneous lumbar diskectomy. In: *Percutaneous Lumbar Diskectomy*, pp. 153–6. Edited by Mayer HM, Brock M. Springer-Verlag, Berlin.

16 Luft C, Weber J, Horvath W *et al.* (1992). Automated percutaneous lumbar discectomy (APLD): Method and 1-year follow-up. *European Radiology* **2**: 292–8.

17 Moison JJF, Battikha M, Madi A *et al.* (1991). Results of a multicentric study of 297 cases automated percutaneous discectomy. *Toft/Hoogland Course*. Munich.

18 Rezaian SM, Silver ML (1989). Percutaneous diskectomy: Personal observations of 27 cases. In: *Percutaneous Lumbar Diskectomy*, pp. 173–6. Edited by Mayer HM, Brock M. Springer-Verlag, Berlin.

19 Solini A, Paschero B, Ruggieri N (1991). Automated percutaneous diskectomy according to the Onik method: Conclusive considerations. *Italian Journal of Orthopaedics and Traumatology* **17**(2): 225–36.

20 Vanneroy F, Courtheoux F, Huet H *et al.* (1991). Percutaneous automated nucleotomy in the treatment of foraminal and extra-foraminal lumbar disk herniation: Review of 18 cases. *Rachis* **3**(4): 323–33

21 Wilson LF, Mulholland RC (1992). Automated percutaneous discectomy versus surgery: A prospective randomised study of treatment for lumbar disc protusions. *3rd Annual Meeting of the European Spine Society*, p. 616. Cambridge, September.

22 Gobin P, Theron J, Courtheoux F *et al.* (1990). Percutaneous automated lumbar nucleotomy. *Journal of Radiology* **71**(6–7): 401–6.

23 Carter H, Menkes CJ, Cormier C *et al.* (1991). Post-nucleolysis low-back pain: Influence of early ambulation. *Toft/Hoogland Course*. Munich.

24 Javid M (1988). Signs and symptoms after chemonucleolysis. *Spine* **13**: 1428–37.

25 Lavyne MH (1992). Complications of percutaneous laser nucleolysis. *Journal of Neurosurgery* **76**: 1041.

26 Bonati AO (1991). Arthroscopic lumbar laser discectomy *Toft/Hoogland Course*. Munich.

27 Cooney F (1991). Comparison of the early results of laser lumbar decompression to chemonucleolysis and automated percutaneous lumbar discectomy. *Toft/Hoogland Course*. Munich.

28 Castro WHM, Jerosch J, Hepp R *et al.* (1992). Restriction of indication for automated percutaneous lumbar discectomy based on computed tomographic discography. *Spine* **17**(10): 1239–43.

29 Weigand H, Weissner B (1989). Percutaneous nucleotomy: A new interventional radiological technique for lumbar disk removal. *Annals of Radiology* **32**(1): 34–7.

30 Seibel RMM, Gronemeyer DHW, Sorensen RAL (1992). Percutaneous nucleotomy with CT and fluoroscopic guidance. *Journal of Vascular and Interventional Radiology* **3**: 571–6.

31 Hammon W (1989). Percutaneous lumbar nucleotomy. *Neurosurgery* **24**(4): 635.

32 Onik GM, Helms CA (1991). Automated percutaneous lumbar diskectomy. *American Journal of Radiology* **156**: 531–8.

33 Bocchi L, Ferrata P, Passarello F *et al.* (1989). La nucleoaspirazione secondo Onik nel trattamento dell' ernia discale lombare analisi multicentre dei primi risultati su oltre 650 trattamenti. *Rivista di Neuroradiologia* **2** (Suppl. 1): 119–22.

34 Schweigel J (1988). Automated percutaneous discectomy: Comparison with chymopapain. In: *Automated Percutaneous Discectomy*, pp. 71–7. Radiology Research and Education Foundation, San Francisco.

35 Ulrich HW (1992). Automated percutaneous diskectomy: Indication, technique and results after 2 years. *Zeitschrift für Orthopädie* **130**(1): 45–50.

36 Sang Keun Chang (1990). Chemonucleolysis vs percutaneous automated nucleotome in lumbar disc surgery. *Konruk University School of Medicine Academic Journal* **34**(2): 43–56.

37 Cooney FD (1989). Comparison of chemonucleolysis with chymopapain to percutaneous automated diskectomy: A surgeon's first 50 cases of each. In: *Percutaneous Lumbar Diskectomy*, pp. 163–8. Edited by Mayer HM, Brock M. Springer-Verlag, Berlin.

38 Lavignolle B, Castagnera L, Grenier N. Comparative study of nucleolysis versus percutaneous nucleotomy versus automated discectomy in treatment of herniated disc. *Nucleotomie Percutanée Manuelle et Nucléoaspiration* 287–95.

39 Revel M, Payan C, Valley C *et al.* (1993). Automated percutaneous lumbar discectomy versus chemonucleolysis in the treatment of sciatica: A randomized multicenter trial. *Spine* **18**(1): 1–7.

40 Swiecicki M (1989). Results of percutaneous lumbar diskectomy compared to laminectomy and chemonucleolysis. In: *Percutaneous Lumbar Diskectomy*, pp. 133–7. Edited by Mayer HM, Brock M. Springer-Verlag, Berlin.

CHAPTER 10

Selective nerve root blocks

Nikolai Bogduk, Charles Aprill and Richard Derby

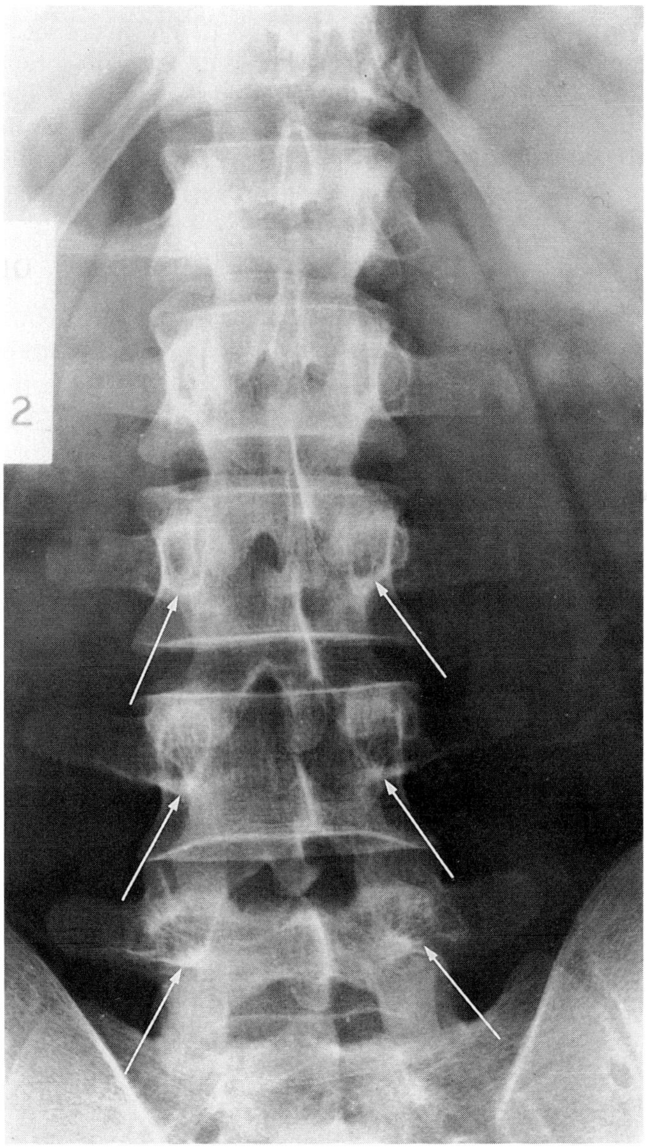

Nerve root blocks were developed as a means by which to define the source, if not the cause, of nerve root pain.[1–10] Their value was purported to lie in patients in whom imaging studies were equivocal or in whom imaging studies suggested possible compression of several lumbar nerve roots.[1,2,6,9] By provoking and anaesthetizing a putatively symptomatic nerve root, a physician could determine whether or not that root was the source of the patient's symptoms. Once the symptomatic root was identified in this way, surgical therapy could be directed selectively at that nerve and not at others which were not symptomatic despite appearing compromised on imaging studies. However, what is referred to as the lumbar 'nerve root' is not a simple structure.

Within each lumbar intervertebral foramen the rootlets of the dorsal and ventral roots converge to form the spinal nerve, with the dorsal root ganglion lying just proximal to the spinal nerve (Fig. 10.1). The nerve roots approach the intervertebral foramen by circumventing the medial aspect of the pedicle immediately above the foramen, lying in the entrance zone of the root canal, otherwise known as the lateral recess. The nerve roots and their accompanying radicular veins and arteries are enclosed in a tapering sleeve of dura mater which peripherally blends with the epineurium of the spinal nerve (Fig. 10.1).

Within the vertebral canal and the intervertebral foramen the dural sleeve is surrounded by loose, fibrous tissue and fat continuous with that surrounding the thecal sac. Ventral to the nerve root sleeve run the sinuvertebral nerves. Typically these are represented by a major filament that passes transversely just below the pedicle to enter the vertebral canal in company with the anterior spinal canal artery.[11] Additionally or alternatively several filaments enter the vertebral canal ventral to the nerve root sleeve. The sinuvertebral nerves innervate the ventral aspect of the thecal sac and nerve root sleeve and furnish branches to the intervertebral discs and posterior longitudinal ligament forming the floor of the vertebral canal.

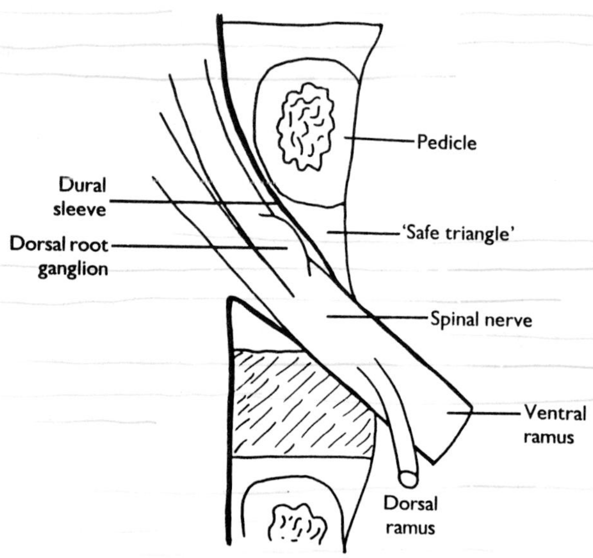

Fig. 10.1
Right lumbar spinal nerve viewed from the rear, with the dural sleeve opened, showing the relationship of the spinal, its roots and its ventral ramus to the pedicle. The 'safe triangle' is the region where a needle may be introduced without striking the neural elements or entering the dural sleeve.

Viewed against this background of anatomy the effect and interpretation of nerve root blocks are not as simple as originally portrayed. Local anaesthetics injected around the nerve root will anaesthetize not only the roots themselves and the spinal nerve but also the dural sleeve and the sinuvertebral nerves. Consequently, although segment specific a nerve root block cannot be tissue specific.

Application

Selective nerve root blocks can be used to test various hypotheses. Since they anaesthetize the nerve roots and spinal nerve they can test whether or not a patient's pain stems from these structures or is mediated by them, but since they also anaesthetize the dura mater they can test the hypothesis that a patient's pain stems from the nerve root sleeve. Moreoever, since they block the sinuvertebral nerves, nerve root blocks test whether or not a patient's pain stems from the discs or other tissues supplied by these nerves. Nerve root blocks, however, cannot discriminate between these various hypotheses. Since all of the above structures are blocked simultaneously, relief of pain does not implicate one source of pain ahead of any other. Any putative discrimination must be based on the patient's clinical features.

If the patient clearly suffers from radicular pain, namely shooting or lancinating pain in the lower limb along a narrow band,[11] relief of that pain by a nerve root block reliably implicates the anaesthetized root as the source of that pain. If, on the other hand, the patient suffers from somatic pain – deep, dull, aching pain in the back or referred into the lower limb[11] – relief of that pain by a nerve root block implicates no particular structure. The pain could stem from any of the structures innervated by the sinuvertebral nerves of that level or from any of the structures innervated by the spinal nerve at that level. The former include the nerve root sleeve, the thecal sac and the intervertebral discs. The latter include discs, muscles and zygapophysial joints.

Partial relief of somatic pain may occur because of the multisegmental innervation of certain structures. For example, the sinuvertebral nerve and the spinal nerve in a particular intervertebral foramen innervate the disc at that level plus the disc of the segment above. Blocking these nerves, therefore, may partially relieve pain from either of these discs. Thus, in the context of somatic pain, a nerve root block is not source specific, and is of value only if it can be considered in the context of the results of discography and medial branch blocks or zygapophyseal joint blocks which might otherwise implicate or exclude the disc or posterior elements as the source of pain.

Technique

Early techniques of lumbar nerve root blocks required that the needle be directed at the target nerve, usually just outside the intervertebral foramen; the objective was to reproduce the patient's pain by striking the nerve.[1,3,6,7,9,10] Concordance between the evoked pain and the patient's accustomed pain was taken as the cardinal indication that the 'impaled' nerve was responsible for the patient's pain. In contemporary practice this aspect is regarded as inappropriate and unnecessary.

Modern techniques avoid deliberately striking the nerve. This renders the procedure more tolerable for the patient and reduces the potential risk of nerve damage. Instead, the diagnostic decision is based on whether or not blocking the target nerve reduces the patient's symptoms. Observations are made opportunistically as to if and when the patient's pain is reproduced, but this is not an essential component of the diagnostic process.

Lumbar nerve root blocks

The most practicable approach for nerve root blocks at lumbar levels involves the target point at the base of the pedicle immediately above the target nerve. Radiographically, the target point lies infero-lateral to the pedicle, i.e. at the 5:30 position on the right and at the 6:30 on the left, using an analogy with a clock-face (Fig. 10.2). This target point lies at the medial apex of what can be portrayed as a 'safe triangle' (Fig. 10.1). The triangle has a base tangential to the pedicle, a side in line with the outer margin of the intervertebral foramen and a hypotenuse coincident with the upper margin of the spinal nerve and dorsal root ganglion. A needle tip directed into this triangle will therefore lie above and lateral to the nerve and will not incur any other structure of significance to risk of morbidity.

To access this target point a 22 G or a 25 G spinal needle is inserted through the skin and back muscles along an oblique approach. The puncture point is determined by obtaining an oblique view of the target intervertebral foramen such that the apex of the superior articular process of the ipsisegmental zygapophyseal joint points directly upwards towards the target pedicle (Fig. 10.3a). The needle is passed through the skin just above and lateral to this apex. Under repeated fluoroscopic screening, the needle is advanced slowly towards the base of the pedicle until its further advance is arrested by bony contact (Fig. 10.3b). At this stage, its tip should be in correct position, which should be confirmed by postero-anterior and lateral views (Fig. 10.4). If not, the tip should be readjusted until it assumes correct position.

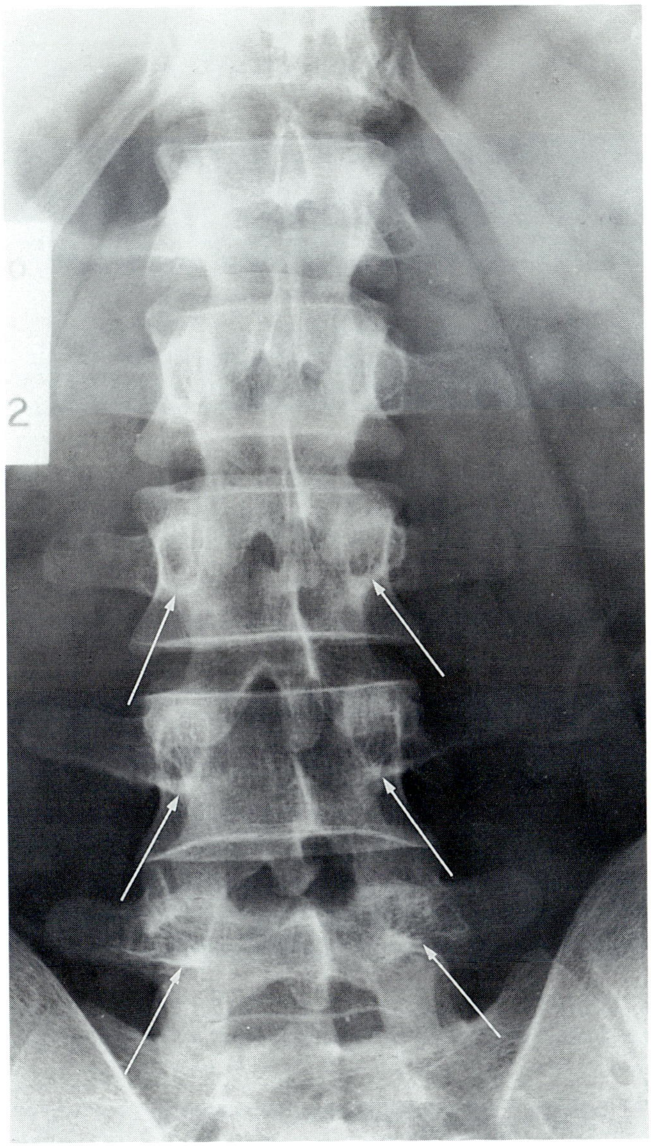

Fig. 10.2
A postero-anterior view of lumbar pedicles showing the target points (arrowed) for selective nerve root blocks.

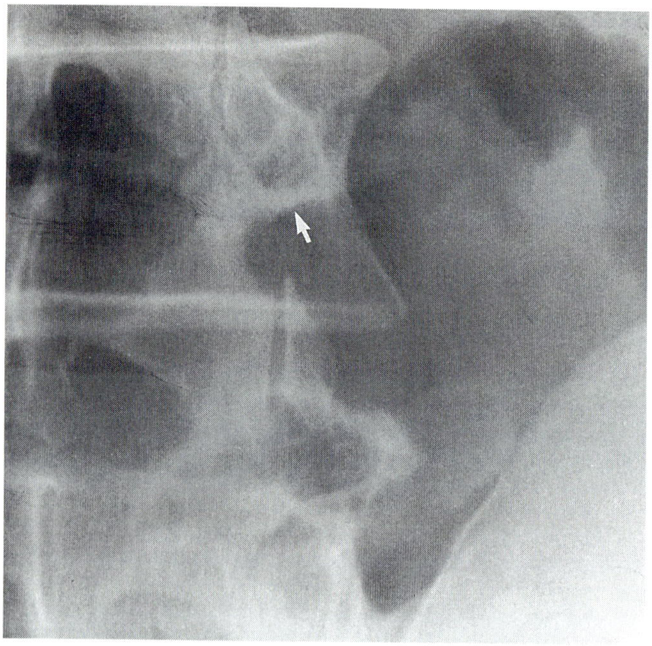

(a)

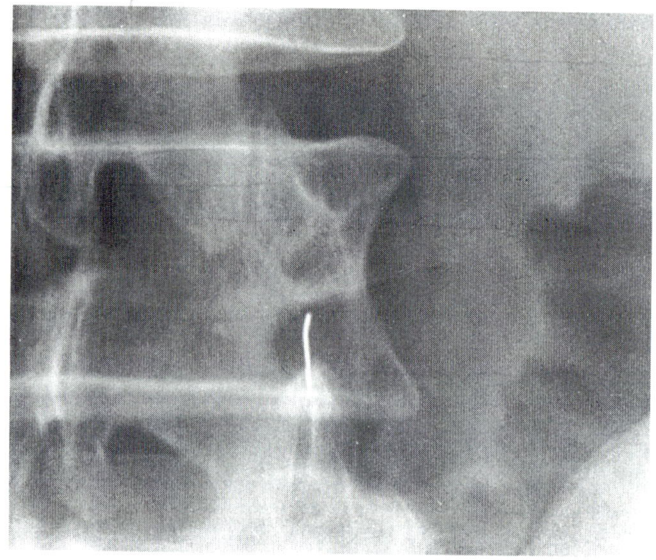

(b)

Fig. 10.3
(a) Oblique view of a right L4–5 intervertebral foramen to illustrate how the superior articular process points towards the target point (arrowed) for a selective nerve root block. (b) A needle in correct position for a selective lumbar nerve root block.

Once the needle is in correct position, two staged injections are made. The second will be an injection of local anaesthetic or local anaesthetic mixed with corticosteroid. The first will be an injection of contrast medium, the purpose of which is to verify correct placement of the needle but also to determine the volume of injectate that can and should be injected to achieve a block without compromising its selectivity.

One millilitre of contrast medium should be injected slowly under direct visualization to indicate the direction and extent of spread of any solutions that might subsequently be injected. An appropriate pattern of spread is one in which the contrast medium flows along the surface of the nerve root complex outlining the bulge of the dorsal root ganglion and the course of the nerve root sleeve (Fig. 10.4). Centrally, the contrast medium spreads ventral to the nerve root sleeve curving upwards and medially around the pedicle and extending medially into the epidural space. Peripherally, the contrast medium outlines the course of the ventral ramus to greater or lesser extents.

A sufficient volume should be injected to outline the target nerve but not more. The contrast medium should not be allowed to reach the next spinal nerve lest the selectivity of the block be jeopardized. Usually about 1.0 ml is sufficient to outline the target nerve; by 2.0 ml the contrast medium starts to reach the nerve next above. Not more than 2.0 ml of contrast medium should be injected unless the flow is predominantly in a peripheral direction. In that event it is better to readjust the position of the needle slightly to achieve a predominantly central dispersal.

As a precaution against the injection of excessive volumes of contrast medium or subsequent solutions, only 2-ml or 3-ml syringes should be used. All injections should be performed slowly at the rate of about 1.0 ml per 20 seconds. The patient should be warned to expect pain during the injection of contrast medium and should be asked to report whether or not the evoked pain is concordant in quality and distribution with the pain they usually suffer.

Failure of the contrast medium to spread centrally may indicate epidural pathology – adhesions, scar, disc herniation or stenosis – but inferences as to the nature of the pathology cannot be drawn reliably on the basis of the epidurogram.[10] Good-quality imaging studies will better define any pathology. If required, post-injection CT scanning can be particularly illuminating in this regard.

During the injection of contrast medium the opportunity is taken to record any pain response offered by the patient. Notes should be taken of the location and appearance of the leading edge of the spreading contrast medium. Pain reproduction can be interpreted in the light of other imaging studies to determine whether it is consistent or not with the location and nature of the lesion perceived to be responsible for the patient's symptoms. For example, pain reproduction early in the course of the injection when the contrast medium is still in the intervertebral foramen could be consistent with

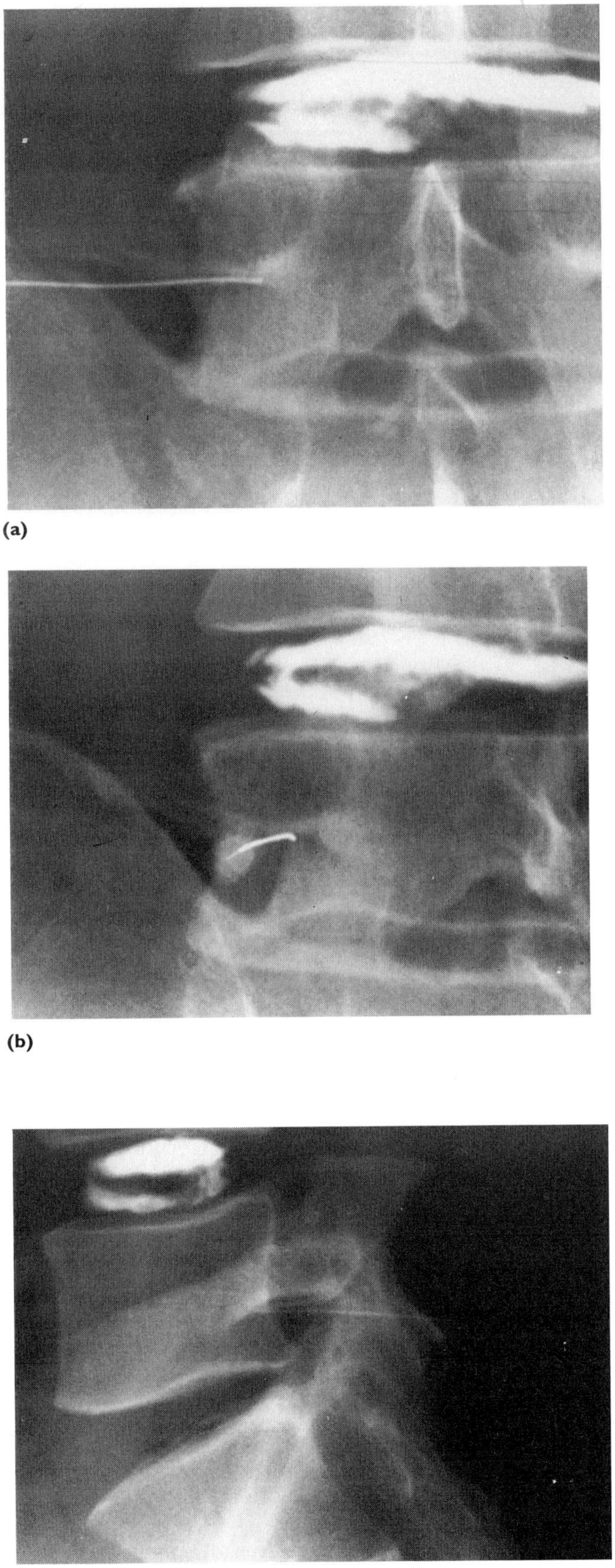

(a)

(b)

(c)

Fig. 10.4
Stages in the execution of a left L5 selective nerve root block. (a), (b), (c) Posterior, oblique and lateral views of a needle in correct position prior to the injection of contrast medium.

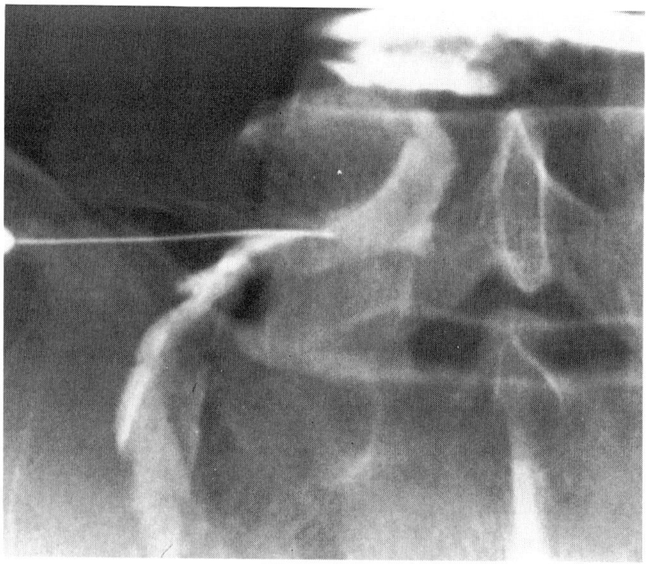

(d)

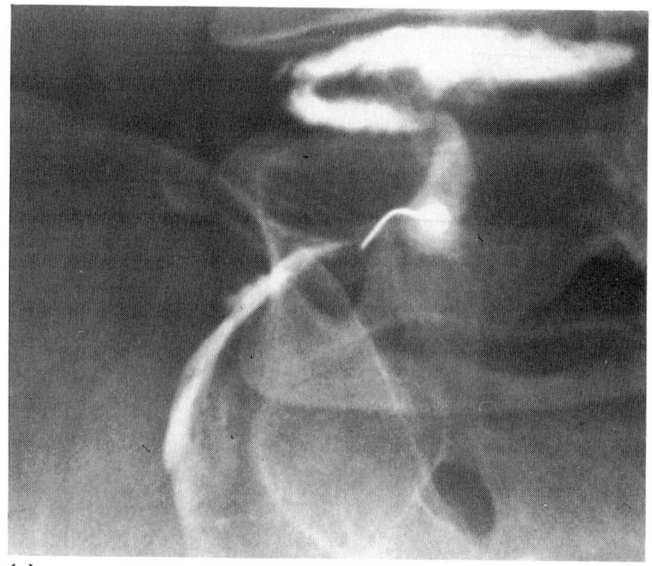

(e)

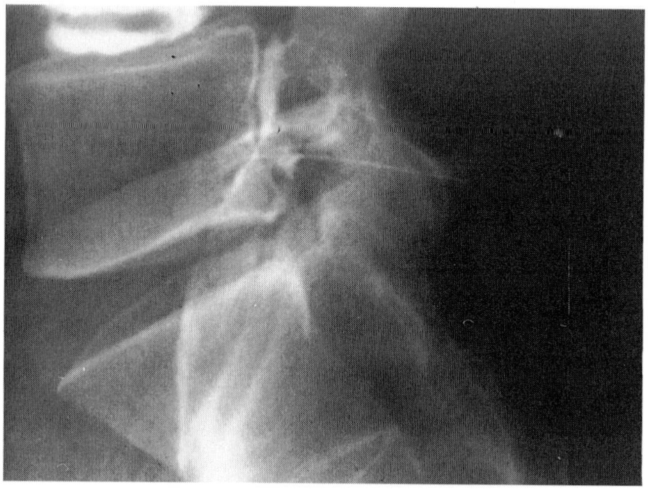

(f)

Fig. 10.4 *Continued*
(d), (e) and (f) Posterior, oblique and lateral views following the injection of 1.5 ml of contrast medium.

foraminal stenosis or a far, lateral disc herniation. Late reproduction of pain when the contrast medium approaches the disc above could be consistent with the sequestrated fragment from that level (Fig. 10.5).

Failure to outline the nerve root complex can occur if the injection is intravascular. To reduce the risk of intravenous injection, the patient must be encouraged to breathe normally and not hold their breath. This minimizes the pressure in the epidural venus plexuses and reduces their distension. Distended epidural veins not only increase the risk of vascular injection but also impede the flow of contrast medium into the epidural and periradicular spaces. If inadvertent intravenous injection does occur its appearance is obvious; the contrast medium dissipates rapidly into the vessels and is cleared from the field; it does not persist and outline the nerve root complex. Should intravenous injection occur, the needle should be readjusted slightly and a renewed injection of contrast medium undertaken.

Once an appropriate dispersal of contrast medium has been established, the syringe containing the contrast medium is replaced with one containing the next agent. This can be local anaesthetic alone, for purely diagnostic purposes, or a mixture of local anaesthetic and corticosteroid for combined diagnostic and putatively therapeutic purposes. Up to 2 ml of agent should be injected at the same site and at the same rate at which the contrast medium had been injected.

When local anaesthetic is used as a sole agent the intention is to anaesthetize the nerve root and its surroundings. Technical success is evident by the onset of numbness in the appropriate dermatome. For this purpose, a long-acting local anaesthetic such as 0.5% or 0.75% bupivacaine is recommended. This provides a prolonged period of

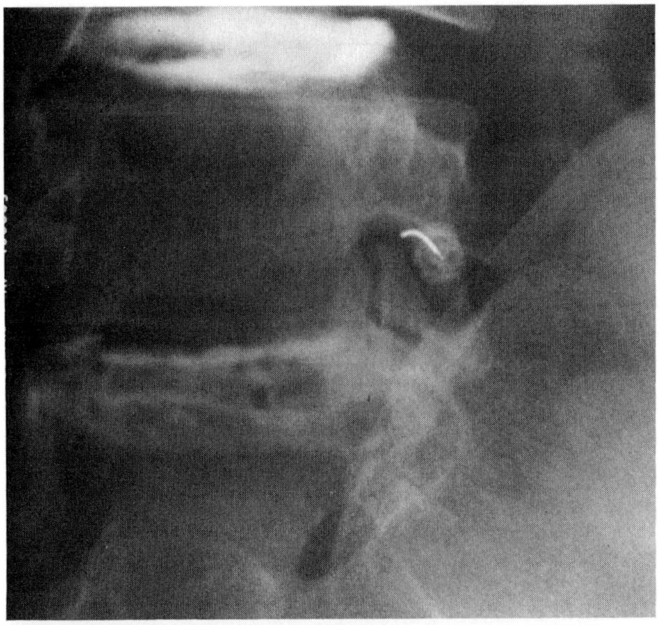

(a)

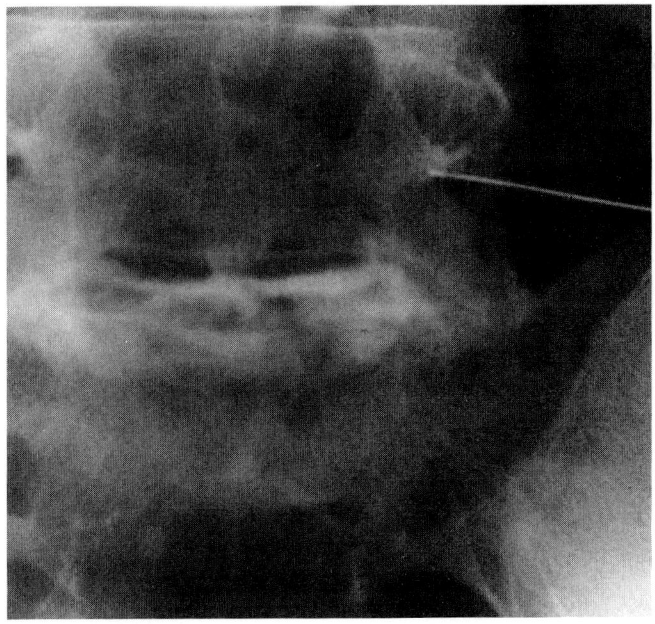

(b)

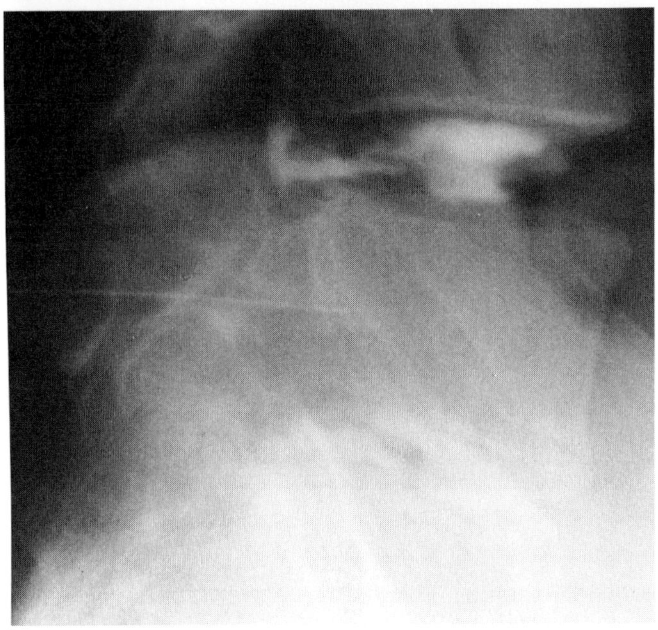

(c)

Fig. 10.5
Stages in the execution of a right L5 selective nerve root block. (a), (b), (c) Oblique, posterior and lateral views of the needle in correct position prior to the injection of contrast medium. In (c) note the posterior leakage of contrast medium following previous discography indicating rupture of the L4–5 disc.

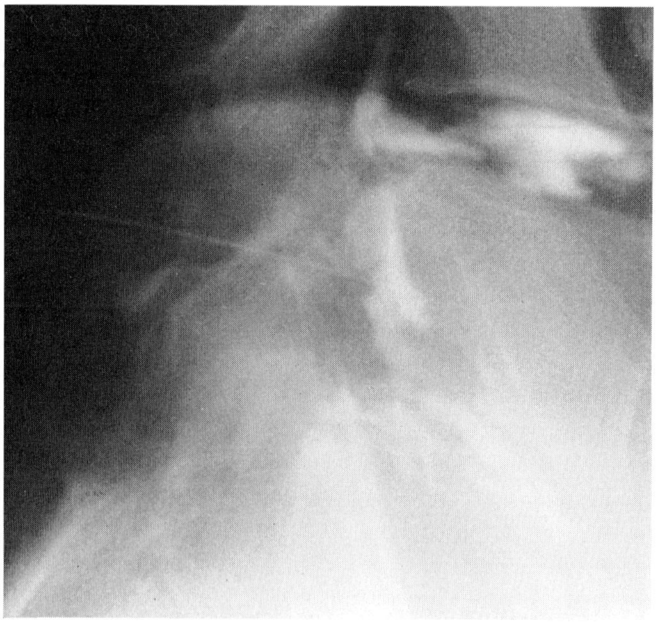

(d)

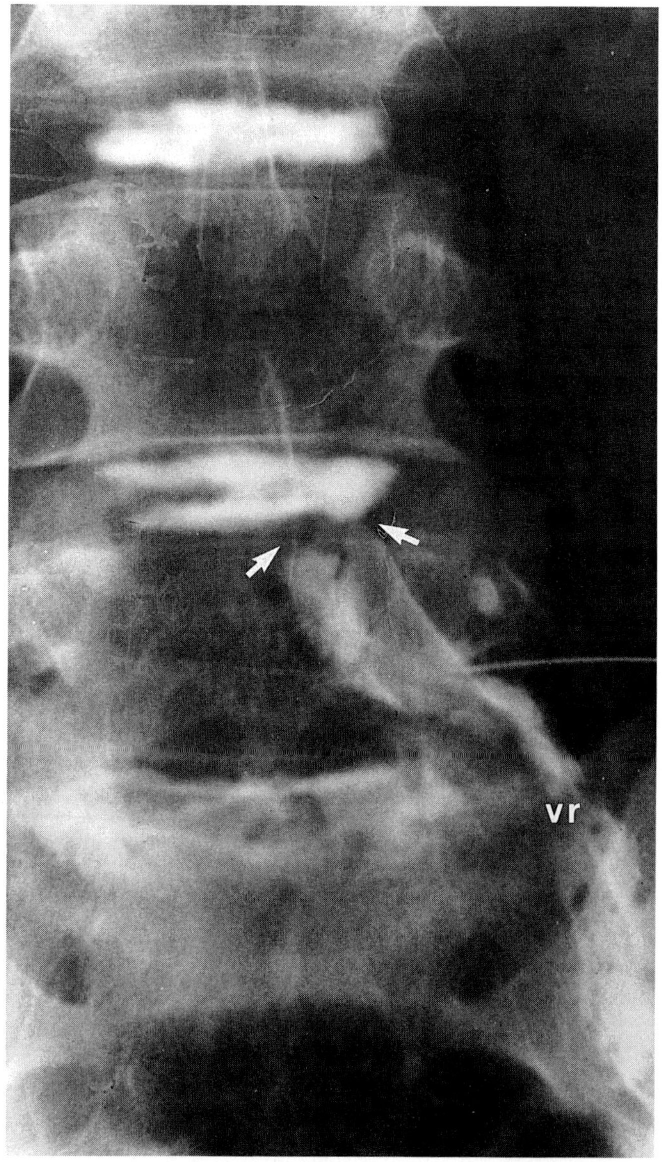

(e)

Fig. 10.5 *Continued*
(d), (e) Lateral and posterior views following the injection of 1.5 ml of constrast medium. In (e) the contrast medium follows the ventral ramus (vr) laterally and inferiorly. Medially it circumvents the pedicle of L5 and extends upwards towards the L4–5 disc. Injection was terminated immediately when the patient reported reproduction of their accustomed radicular pain which coincided with the advancing front of the contrast medium (arrows) reaching the caudal edge of the ruptured L4–5 disc ((d) and (e); compare with (c)).

anaesthesia during which the patient can evaluate the effect on their symptoms.

The same local anaesthetic may be mixed with a corticosteroid preparation in equal parts. The objective is to obtain a more prolonged response. The local anaesthetic component provides an immediate diagnostic effect while the corticosteroid is intended to provide a more sustained, quasi-therapeutic effect.

S1 nerve root blocks

The technique for blocking the S1 nerve root is governed by the different anatomy of the sacrum and its foramina. In a patient lying prone, the sacrum is typically inclined so that the posterior and anterior sacral foramina are not coincident along postero-anterior views. If desired, and if C-arm fluoroscopy is available, the X-ray tube can be tilted in a cephalo-caudad direction along the length of the patient to bring the posterior and anterior sacral foramina into view in a coincident pattern, but this is not essential.

The S1 nerve roots course medial to the S1 pedicle before leaving the sacrum through the S1 anterior sacral foramen which lies below and lateral to the pedicle. The target point for an S1 block lies at the inferior medial corner of the pedicle and access to this point is obtained through the posterior sacral foramen (Fig. 10.6). On postero-anterior screening, what should be visualized is the S1 pedicle.

A 25 G or 22 G spinal needle should be inserted through the skin behind the sacrum slightly lateral and below the target point on the S1 pedicle so that the

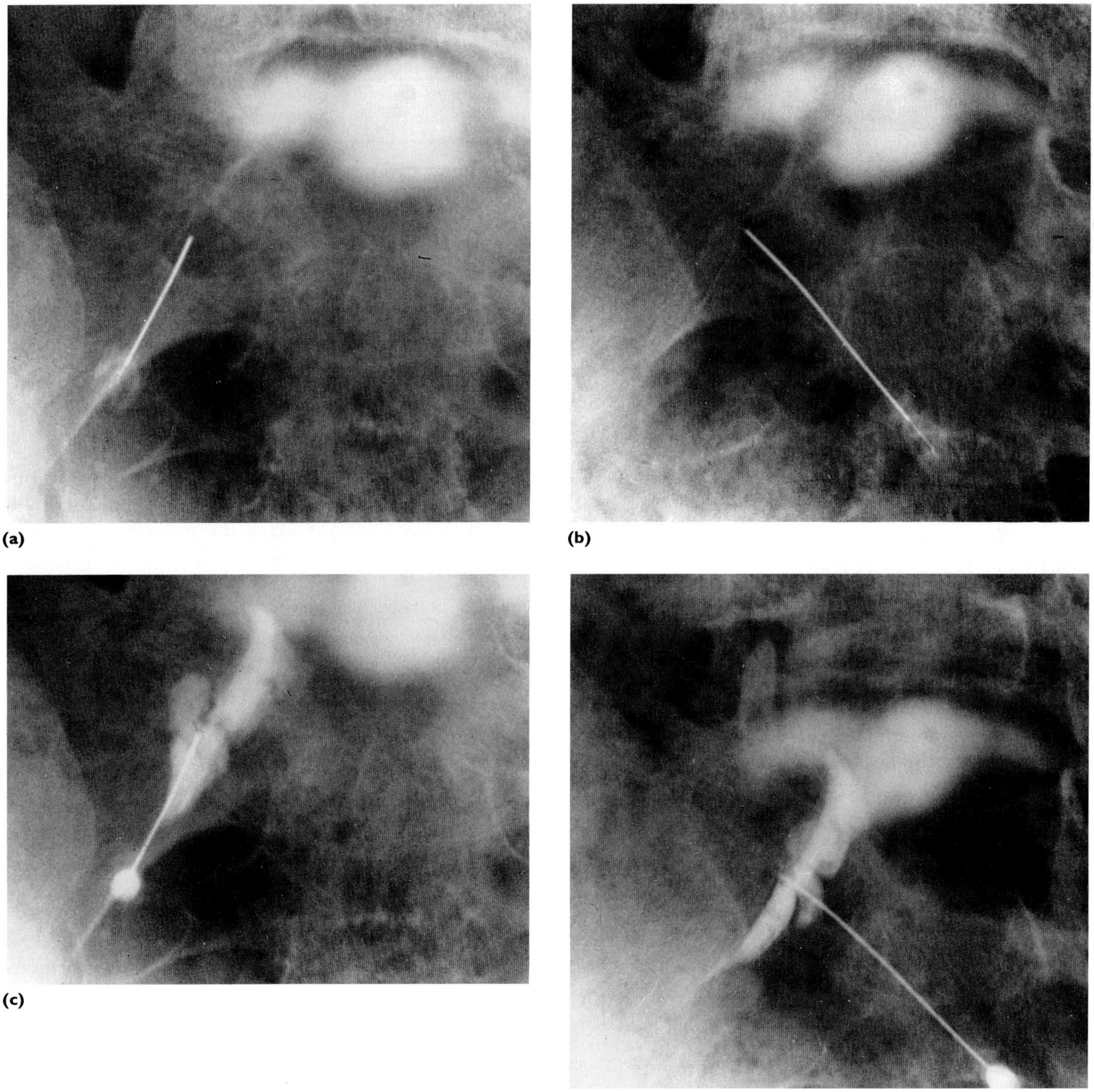

(a)

(b)

(c)

(d)

Fig. 10.6
Stages in the execution of a left S1 selective nerve root block. (a) Posterior and (b) oblique view of a needle in correct position on the target point. (c) Posterior and (d) oblique view after injection of 1.0 ml of contrast medium.

needle passes towards the target point with a slight medial and cephalad orientation. The objective is to have the tip of the needle rest on the medial end of the caudal surface of the pedicle behind the anterior wall of the sacrum. To achieve this position, the needle must pass through the posterior sacral foramen but must not leave the sacrum through the anterior sacral foramen.

If the posterior sacral foramen can be visualized its margin can be negotiated under direct vision. If the posterior sacral foramen cannot be visualized it can none the less be negotiated by 'feel'. By aiming the needle cephalad of the target point, after penetrating the skin, erector spinae aponeurosis and multifidus muscle, the tip will strike the dorsal surface of the sacrum above the S1 posterior sacral foramen. Thereafter, to enter the foramen, the needle need only be readjusted progressively caudad so that it essentially 'walks' down the superior wall of the foramen which is formed by the S1 pedicle. Success in this manoeuvre will be indicated by progressive increases in the depth of penetration of the needle until it arrives at the target point.

Passage through the anterior sacral foramen is avoided by maintaining contact with the S1 pedicle and by maintaining a medial orientation of the needle so that it is inclined towards the sacral canal. Passage through the anterior sacral foramen will be indicated by loss of resistance, in which case the needle should be withdrawn and replaced in contact with the pedicle.

Once the needle is in position, contrast medium and subsequent agents can be injected following the same protocol as for lumbar nerve root blocks.

Interpretation

The archetypical positive response to a selective nerve root block is that once the root is anaesthetized the patient is relieved of their pain. Such a clear response indicates unequivocally that the root is either the source of the patient's pain or is the sole segmental nerve that is mediating the patient's pain. No relief of pain excludes the root as either the source of pain or the pathway by which the pain is mediated.

More commonly, incomplete or mixed patterns of response occur. The patient may be relieved of their radicular pain but continues to complain of low back pain or referred pain. Such a response implicates the anaesthetized root as the source of their radicular pain but not their somatic pain.

The patient may be partially relieved of their radicular pain, in which case another root may be contributing to their radicular pain. Blocking a root at a second level may succeed in totally abolishing their pain. Such a response implicates both roots.

Most vexatious is a response in which the patient is partially relieved of their somatic pain. Such a response indicates that one of the structures innervated by that segment is contributing to the patient's pain. Partial relief may occur either because other structures are also contributing to the patient's overall pain or because the one structure that is the sole cause of their pain has only partially been anaesthetized by the root block. These possibilities cannot be resolved by root blocks alone. Patients with this pattern of response should be re-evaluated by other means. Discography can be used to evaluate the disc at the level in question; zygapophyseal joint blocks can be used to evaluate zygapophyseal joints.

In a patient with somatic pain and in whom discography and zygapophyseal joint blocks are both negative, a positive response to a selective nerve root block strongly suggests that the dural sleeve is the source of their pain. Unfortunately, there are no more definitive means of confirming this contention; selective nerve root blocks applied in this way are the only means by which some degree of confirmation may be obtained for a hypothesis that the patient's pain stems from the nerve root sleeve.

Complications

Selective nerve root blocks are not associated with complications provided that they are performed carefully and accurately. The only true, but remote, complications are those of infection and allergic reactions associated with any procedure involving the injection of contrast medium or local anaesthetic.

The foremost hazard of selective nerve root blocks is striking the target nerve, but this is avoided by approaching the target point carefully and accurately. The spinal nerve or dorsal root ganglion can be incurred only if the needle strays caudally. To guard against misadventure, the needle should be introduced slowly and under repeated fluoroscopic monitoring. Under these conditions, should the needle tip strike the nerve it can be promptly withdrawn before it pierces the nerve; thereafter its course can be corrected. This way, the risk of damaging the nerve is reduced or eliminated.

These precautions also guard against inadvertent dural puncture. Provided the needle is directed into the 'safe triangle', it should not pierce the dural sleeve. Furthermore, the injection of contrast medium should reveal whether or not the dura has been pierced. The demonstration of intrathecal spread of contrast medium warns against injecting subsequent agents into the subarachnoid space and calls for repositioning of the needle.

Results

The value of selective nerve root blocks has been defended largely by assertion. Proponents have claimed that nerve root blocks are invaluable in refining a diagnosis of nerve root compression when clinical examination and imaging studies are equivocal.[1,2,6,9] Root blocks are not indicated in every patient with suspected radicular pain. They are indicated only in those cases where doubt exists about exactly which root is the source of pain. Examples include patients in whom myelograms are indeterminate, patients with abnormal segmentation, and patients in whom CT scans or myelograms suggest that more than one root is affected by disc herniation or foraminal stenosis.[9,10] According to some studies, such doubt pertains in up to 20 per cent of patients presenting with apparent radicular pain.[5]

Other illuminating features revealed by the use of selective nerve root blocks are that in the majority of patients, single roots are responsible for their symptoms regardless of pathology.[5] Even in patients with intermittent claudication, symptoms are abolished in 90 per cent of cases by blocking a single nerve.[5]

Occasionally one can encounter prolonged 'therapeutic' effects following selective nerve root blocks in that the patient remains relieved of their pain for more than 6 months.[5] In such cases, surgery can be avoided or deferred.

Why some patients should obtain prolonged relief is an enigma. It could be related to effects of local anaesthetic such as prolonged dampening of C-fibre activity;[12] it could be a physical effect such as clearing 'adhesions' or inflammatory exudates from around the nerve root sleeve; but no data are available by which these possibilities might be distinguished or validated.

When corticosteroids are used two possibilities arise. First, corticosteroids have a prolonged local anaesthetic effect,[13] so that pain relief could be due to this effect rather than an anti-inflammatory action. On the other hand, there is considerable, circumstantial evidence that many cases of lumbar radiculopathy involve inflammation of the nerve root or its adnexae.[14–16] Under these circumstances, the injection of corticosteroids could be perceived as offering a prolonged anti-inflammatory effect to suppress the actual cause of pain from a nerve root.

The implicit value of selective nerve root blocks is that by enabling more accurate diagnosis they lead to better treatment. This contention, however, is largely axiomatic or a matter of faith, for it has never been formally validated. To do so would require comparing the therapeutic outcome of similar cohorts of patients who underwent the same therapy either with or without the benefit of selective nerve root blocks. Such studies have not been performed and, indeed, might be difficult to justify on ethical grounds. The best available data are simply reports that the majority of patients with positive root blocks fare well after surgery, provided the pathology is a disc herniation, bony entrapment of the nerve or periradicular adhesions.[1,6,7,9] Patients with intraneural fibrosis, such as postoperative arachnoiditis, may respond to blocks but respond poorly to those surgical interventions that have been tried.[7,9]

A related utility of selective nerve root blocks is their negative predictive value. The question that arises is: in patients with a positive response to selective nerve root blocks, will surgery relieve their pain? This question has not been addressed using only the immediate, diagnostic response to local anaesthetics; but it has been answered by considering the prolonged response attributed to steroids.

One study examined the surgical outcome of 71 patients who obtained greater than 80 per cent immediate relief of their radicular pain following a diagnostic selective nerve root block.[16] What was correlated subsequently was the outcome of surgical therapy versus the degree of pain relief at 1 week following the block – referred to as the 'steroid response'.[16]

It emerged that, overall, the 'steroid response' offered limited sensitivity but attractive positive and negative predictive values for surgical outcome (Table 10.1). Furthermore, it emerged that this test was of limited value in patients with symptoms of less than 1 year's duration (Table 10.2) but had good negative predictive power in patients with longer durations of symptoms (Table 10.3). In short, patients who fail to obtain sustained relief of radicular pain (relief attributed by inference to the use of steroids) are unlikely to benefit from surgery.

Steroid response	Surgical outcome		
	Positive	Negative	Total
Positive	20	2	22
Negative	11	38	49
Total	31	40	71

Table 10.1 *Correlation between surgical outcome and steroid response in 71 patients who underwent selective nerve root blocks*[16]

Steroid response	Surgical outcome		
	Positive	Negative	Total
Positive	9	0	9
Negative	9	2	11
Total	18	2	20

Table 10.2 *Correlation between surgical outcome and steroid response in 20 patients with leg pain lasting less than 1 year who underwent selective nerve root blocks*[16]

Steroid response	Surgical outcome		
	Positive	Negative	Total
Positive	11	2	13
Negative	2	36	38
Total	13	38	51

Table 10.3 *Correlation between surgical outcome and steroid response in 51 patients with leg pain lasting longer than 1 year who underwent selective nerve root blocks*[16]

Thus, whatever theoretical or axiomatic attraction selective nerve root blocks might have as a diagnostic test to determine whether or not a given root is symptomatic, their further utility is that if corticosteroids are administered in addition to local anaesthetic, they can predict if a patient will not respond to surgery. Using blocks in this way may or may not improve therapeutic outcome but can significantly curtail unnecessary and futile major surgery.

References

1 Krempen JF, Smith BS (1974). Nerve root injection: a method for evaluating the etiology of sciatica. *Journal of Bone and Joint Surgery* **56A**: 1435–44.
2 Krempen JF, Smith BS, De Freest LJ (1975). Selective nerve root infiltration for the evaluation of sciatica. *Orthopedic Clinics of North America* **6**: 311–15.
3 Tajima T, Furukawa K, Kuramochi E (1980). Selective lumbosacral radiculography and block. *Spine* **5**: 68–77.
4 White AH (1983). Injection techniques for the diagnosis and treatment of low back pain. *Orthopedic Clinics of North America* **14**: 553–67.
5 Kikuchi S, Hasue M, Ito T (1984). Anatomic and clinical studies of radicular symptoms. *Spine* **9**: 23–30.
6 Havelsen DC, Smith BS, Myers SR, Pryce ML (1985). The diagnostic accuracy of spinal nerve injection studies: Their role in the evaluation of recurrent sciatica. *Clinical Orthopaedics and Related Research* **198**: 179–83.
7 Dooley JF, McBroom RJ, Taguchi T, McNab I (1988). Nerve root infiltration in the diagnosis of radicular pain. *Spine* **13**: 79–83.
8 Kikuchi S, Hasue M (1988). Combined contrast studies in lumbar spine diseases: Myelography (peridurography) and nerve root infiltration. *Spine* **13**: 1327–31.
9 Herron LD (1989). Selective nerve root block in patient selection for lumbar surgery: Surgical results. *Journal of Spinal Disorders* **2**: 75–9.
10 Stanley D, McLaren MI, Euinton HA, Getty CJM (1990). A prospective study of nerve root infiltration in the diagnosis of sciatica: A comparison with radiculography, computed tomography, and operative findings. *Spine* **15**: 540–3.
11 Bogduk N, Twomey LT (1991). *Clinical Anatomy of the Lumbar Spine*, 2nd edn. Churchill Livingstone, Melbourne.
12 Fink BR, Cairns AM (1987). Differential use dependent (frequency-dependent) effects in single mammalian axons: Data and clinical implications. *Anesthesiology* **67**: 477–84.
13 Johansson A, Hao J, Sjolund B (1990). Local corticosteroid application blocks transmission in normal nociceptive C-fibres. *Acta Anaesthesiologica Scandinavica* **34**, 335–8.
14 Saal JS, Franson RC, Dobrow R, Saal JA, White AH, Goldthwaite N (1990). High levels of inflammatory phospholipase A2 activity in lumbar disc herniation. *Spine* **15**: 674–8.
15 Franson RC, Saal JS, Saal JF (1992). Human disc phospholipase A2 is inflammatory. *Spine* **17**: S129–132.
16 Derby R, Kine G, Saal J, Reynolds J, Goldthwaite N, White A, Hsu K, Zucherman J (1992). Response to steroid and duration of radicular pain as predictors of surgical outcome. *Spine* **17**: S176–S183.

CHAPTER 11

Percutaneous vertebroplasty

Hervé Deramond, Claude Depriester and Pierre Galibert

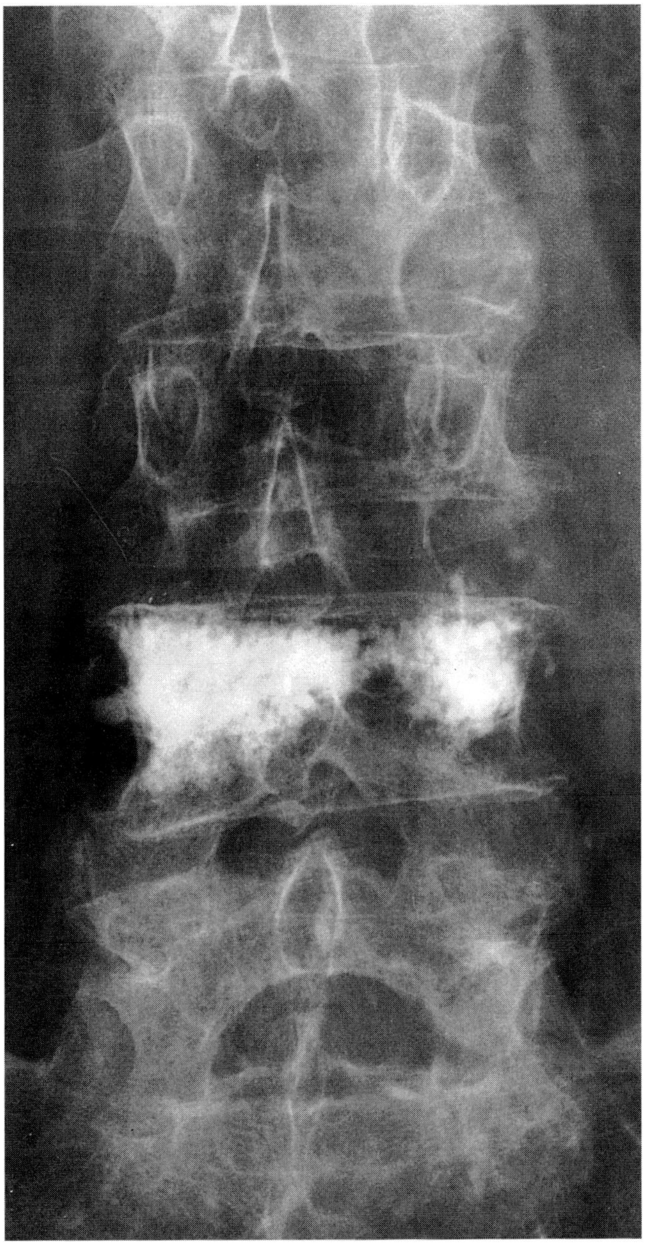

Many lesions of the vertebral body can be treated by a percutaneous approach. The aims of the therapy may be palliative or curative. Percutaneous vertebroplasty with methyl methacrylate (PVMM) is indicated in pathologies that fragment the vertebral body with or without resulting spinal pain.[1-3] The three major indications are aggressive vertebral haemangiomas, osteoporotic vertebral crush fractures and spinal tumours.

This chapter describes percutaneous vertebroplasty and considers how other drugs may be usefully injected into vertebral lesions.

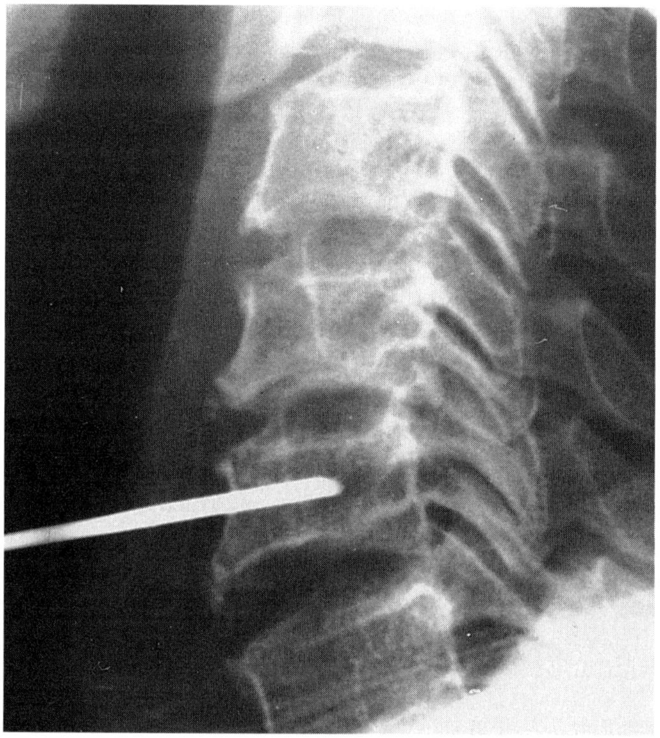

Fig. 11.2
C5 antero-lateral puncture

Technique

The patient should be heavily sedated or anaesthetized. In the sedated patient, local anaesthesia is employed. Fluoroscopic C-arm or biplane guidance is essential.

A 7–15-cm 10–15 G needle is used to puncture the vertebra and to introduce cement. In the lumbar region a 15-cm 10 G needle is employed whilst a shorter (10 cm) and sometimes narrower (15 G) needle is used in the high dorsal or cervical regions (Fig. 11.1).

The approach to the vertebral body varies with the level: antero-lateral in the cervical spine (Fig. 11.2): postero-lateral (Fig. 11.3) or postero transpedicular in the thoracic and lumbar spine (Fig. 11.4). The needle is introduced under fluoroscopic control. Once the periosteum is reached, penetration of the spine is best achieved by lightly tapping the needle with a hammer. This establishes a small portal of entry without fragmentation of bone. A biopsy can precede the cement injection using a trephine

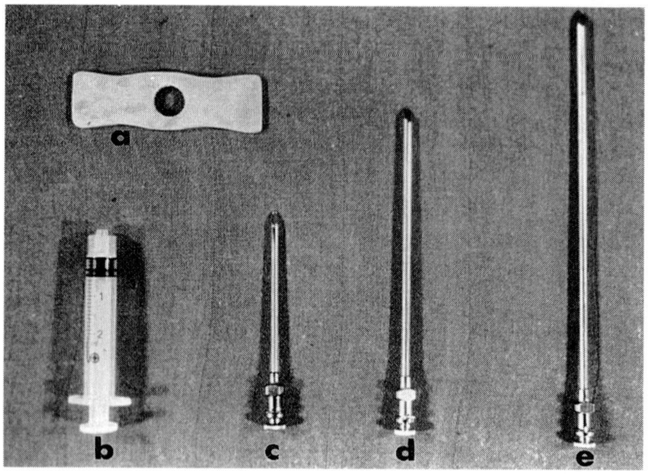

Fig. 11.1 *Material*
(a) Syringe handle to enable efficient injection of the cement. (b) 2.5 ml Luer lock syringe; (c) 7-cm 15 G needle for the cervical spine; (d) 10-cm 10 G needle for the dorsal spine; (e) 15-cm 10 G needle for the lumbar spine.

or a biopsy needle of 15–18 G placed through the vertebroplasty needle (Figs 11.3 and 11.4). When the vertebral lesion is highly vascular, phlebography may be used to identify the perivertebral venous drainage. However, even if there is early venous drainage without vertebral body filling, the injection of acrylic cement is not contraindicated. The cement is prepared: 1 g of tantalum powder, 4 volumes (20 ml) of methyl methacrylate (MM) powder and 1 volume (5 ml) of MM monomer liquid are mixed until the cement has the consistancy of toothpaste.

The mixture is then transferred into Luer lock syringes (2.5 ml) (Fig. 11.1) and pushed into the vertebral body manually under continuous lateral fluoroscopic control, thus avoiding undesirable passage of MM into the spinal canal.[4] Spread into adjacent discs, whilst not contributing to the therapy, does not seem to have any adverse effect. In vascular lesions, prior arterial embolization is not necessary. It is usually necessary to inject each side of the vertebrae. A single puncture allows injection of half the vertebrae; the contralateral puncture fills the remaining space (Fig. 11.5). To avoid adherence of the needle, it should be rotated during the later stages of the injection. Two or three vertebrae may be treated during the same procedure (Fig. 11.6).

For the treatment of complex lesions, or lesions situated at the level of the upper part of the thoracic spine (T1 to T4), puncture may be perfomed under CT

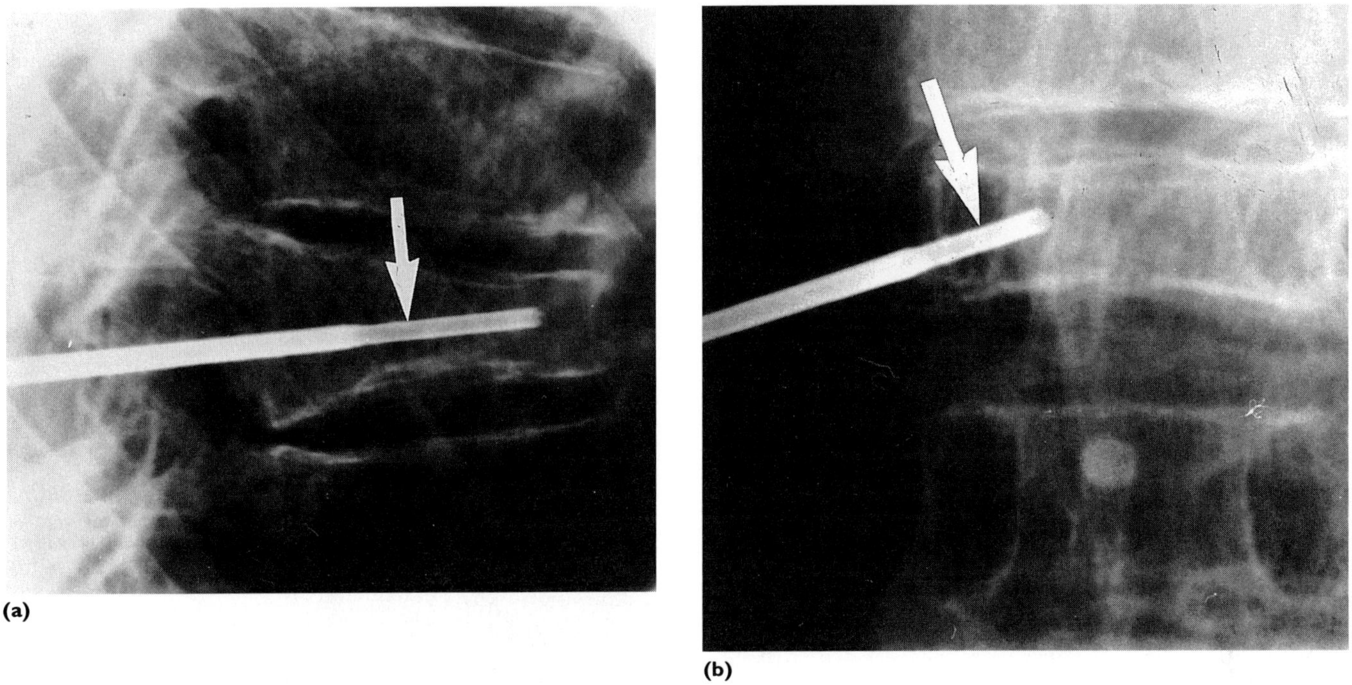

Fig. 11.3
Postero-lateral approach with a biopsy needle (white arrows) placed coaxially through the puncture needle: (a) sagittal and (b) frontal views.

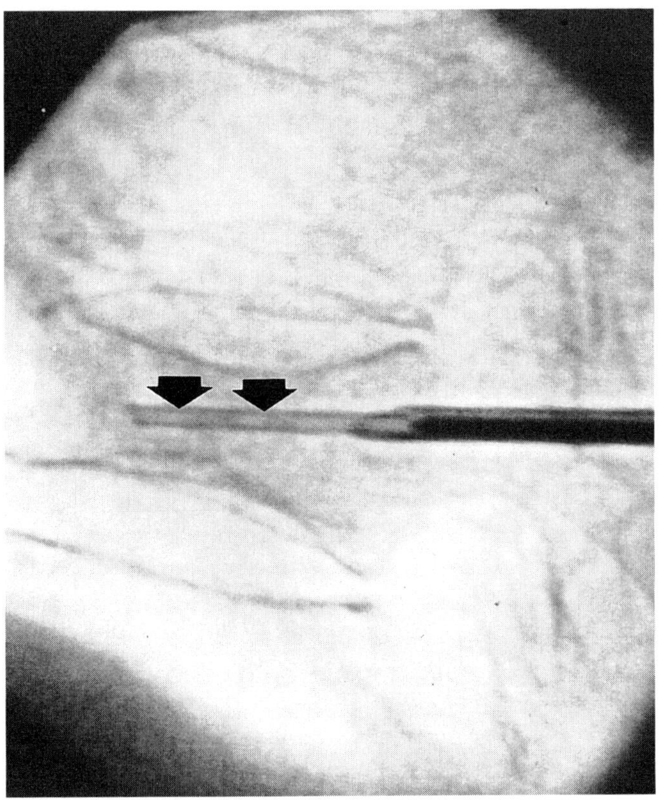

Fig. 11.4
T11 transpedicular approach with a biopsy needle (black arrows).

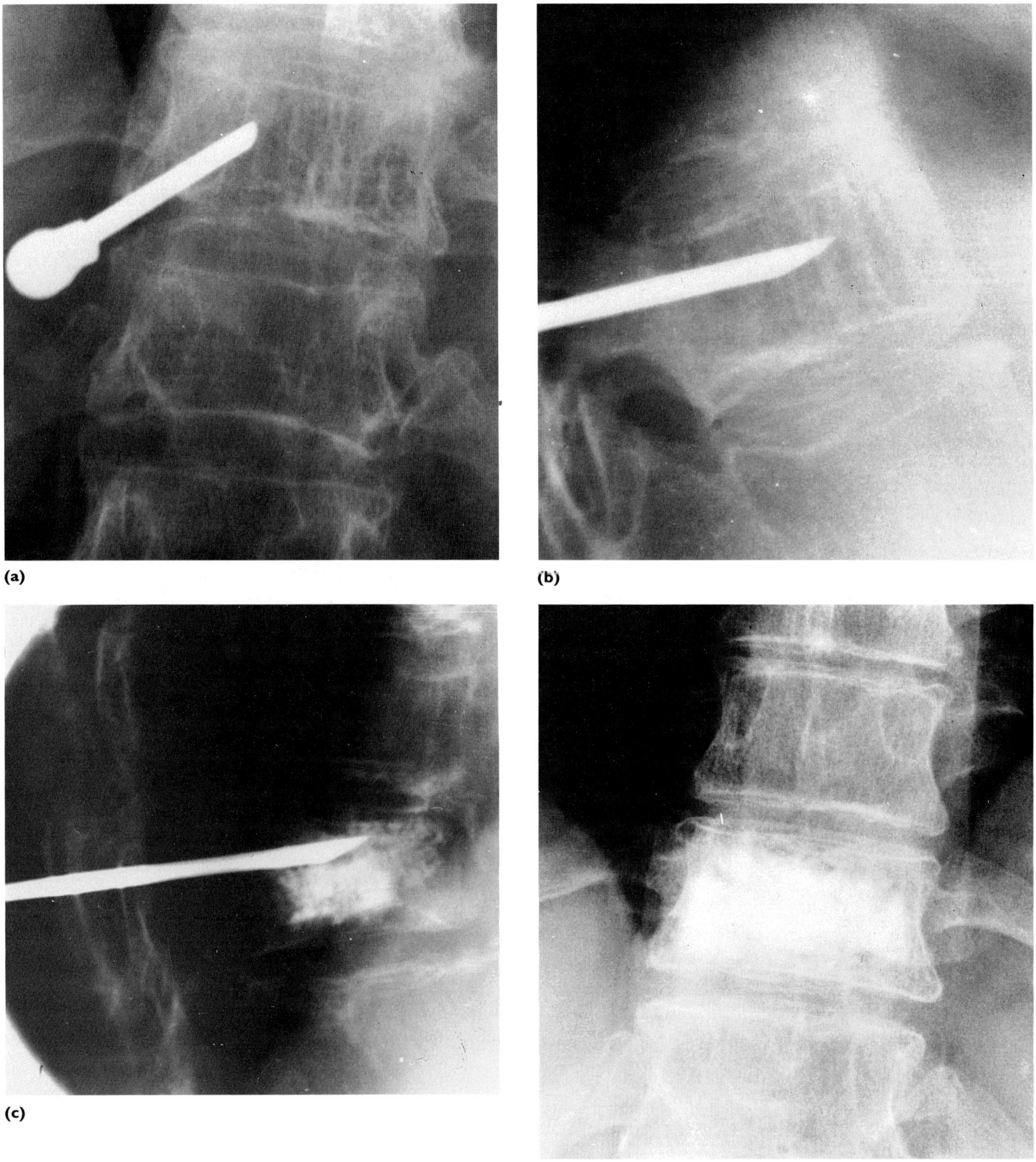

(a) (b) (c) (d)

Fig. 11.5
T11 haemangioma with intense vertebral pain. Right, transpedicular approach (a) sagittal and (b) frontal views. A contralateral puncture is necessary (c) to inject cement in throughout the vertebra (d).

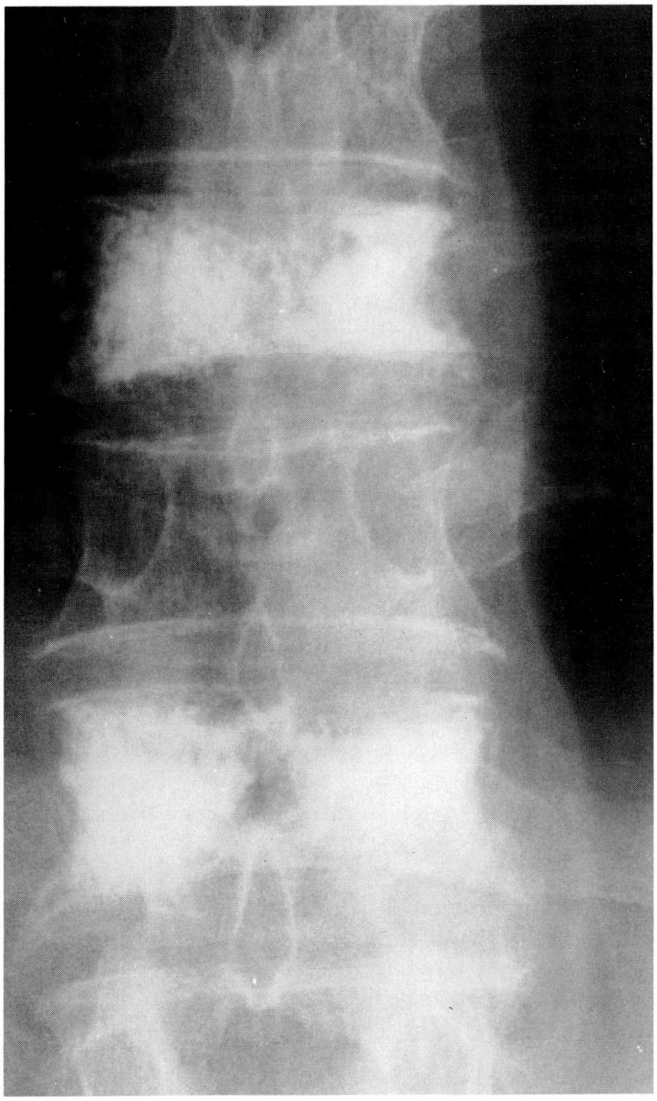

Fig. 11.6
Two vertebrae collapsed owing to osteoporosis have been treated during the same procedure using a bilateral approach at each level.

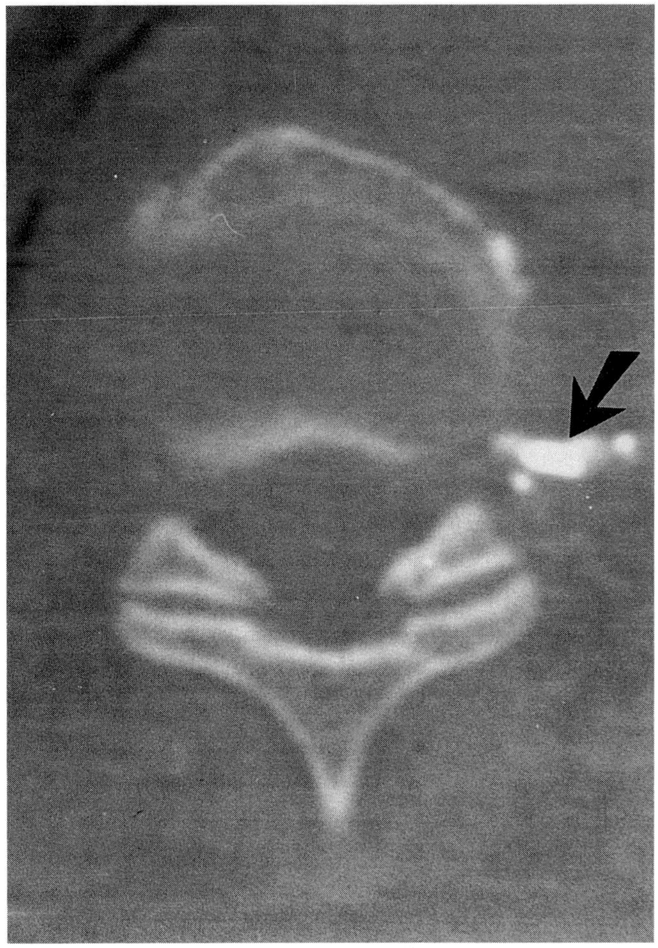

Fig. 11.7
Left T10 leakage of cement (black arrow) inducing a neuralgia which was cured by alcohol neurolysis.

control; in that case a portable C-arm fluoroscopic unit is necessary to allow good lateral fluoroscopic control during the injection of cement.

Incidents and complications

The risks of this method are infection, cord compression and onset or aggravation of radicular pain. To avoid infection it is essential to apply rigorously aseptic conditions, especially in the case of malignant spinal tumours, when chemotherapy frequently diminishes natural defence reactions (1 case in our series).

Spinal cord compression may occur in spinal tumours. The risk increases if there is a rupture of the posterior wall with an epidural or foraminal spread of the tumour. In these cases the indications for percutaneous vertebroplasty must be viewed with great caution. (We have seen 1 case of aggravation of spinal cord compression after vertebroplasty.)

The onset or aggravation of radicular pain is due to leakage of cement around the nerve route in the foramina or in its extra foraminal course (Fig. 11.7). In order to avoid this problem (3 cases in our experience) a transpedicular approach is recommended.

In some cases we have noted an increase in root pain immediately after vertebroplasty. Usually this pain resolves after the administration of anti-inflammatory drugs (ketoprofen IV, 100–300 mg daily).

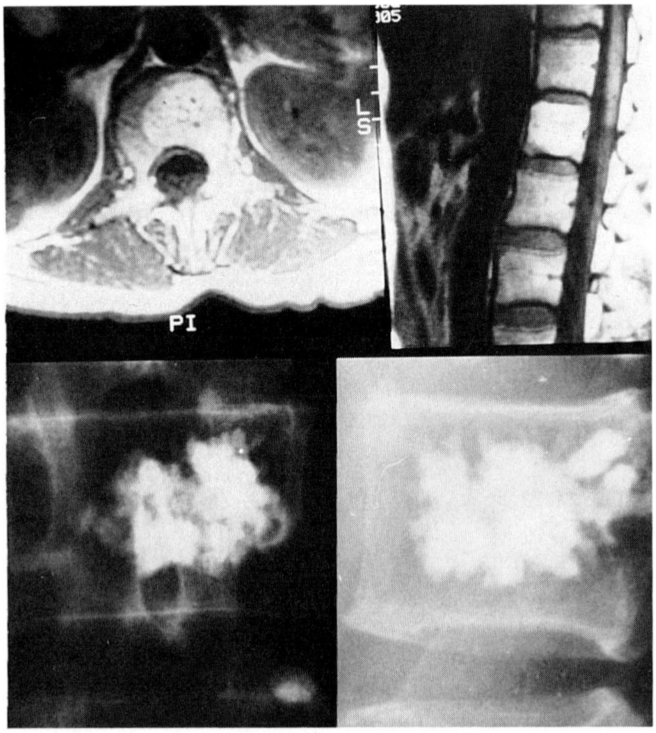

Fig. 11.8 *L2 haemangioma with fat content*
MRI (T1 weighted) before vertebroplasty; and plain films after vertebroplasty.

Indications

PVMM has two objectives: mechanical consolidation and pain relief. The primary indication is the treatment of disease that fragments the vertebral body with or without pain. There are three major indications:

1 vertrebral haemangiomas
2 spinal osteoporosis crush fractures
3 aggressive spinal tumours.

Haemangiomas

Spinal vertebral haemangiomas are commonly asymptomatic. They are benign vascular dysplasias with characteristic radiological findings: vertical trabeculation on plain films; spontaneous decreased density on CT scan; and increased signal on T1 weighted MR images due to their fat content.[8,12] Rare cases present with clinical signs and/or aggressive radiological manifestations: severe spinal pain, spinal cord or nerve root compression;[13,14] thoracic localization; paravertebral and/or epidural extension; unchanged or increased density on CT scan; decreased signal on T1 weighted MR images; and evolution of the lesion (Fig. 11.8).

Vertebroplasty is indicated in vertebral haemangiomas accompanied by severe spinal pain and/or aggressive radiological manifestations without spinal cord or nerve root compression. In aggressive haemangiomas with epidural extension and progressive spinal cord compression the treatment is undertaken in three stages. The first step consists of PVMM, taking care not to inject the epidural component. In the same session, using CT control, the affected posterior arch or the remaining part of the vertebral body not injected with cement is punctured. This allows a complementary injection with a mixture of lipiodol and tissue adhesive to complete the embolization of the dysplasia and its epidural component before laminectomy (Fig.11.9). We have treated more than 70 patients with this disorder. Among them, four underwent open surgery for epidural extension of haemangioma. Most of the patients had intense spinal pain. For the 2 patients without pain, PVMM was employed to achieve mechanical stability. In all but 1 case, vertebroplasty achieved its goal: either consolidation or analgesia. Pain relief was long lasting in all cases (up to 8 years). Radiological results indicated stability in all cases (up to 8 years).

Two incidents were noted: intercostal neuralgias healed after local infiltration. Consolidation and suppression of vertebral mobility are thought to be the mechanism of pain relief. Evolution of the haemangioma is not arrested but if it remains localized to the vertebra, PVMM inhibits extension of the lesion. PVMM can be used in conjunction with other therapeutic techniques, such as surgery or embolization. In our view radiotherapy should be avoided.

Osteoporotic vertebral crush fractures

Vertebral fractures, whether spontaneous or due to minor trauma, are typical complications of osteoporosis. They are the result of mechanical failure of demineralized vertebrae.[15] These crush fractures occur most frequently in post-menopausal women aged 60 years or older.[16–18]

The medical treatment of osteoporosis is aimed towards the prevention of vertebral fracture and uses antiresorptive drugs (oestrogen and calcitonin) or substances such as fluoride that stimulate bone formation.[19–23] These fractures are painful, and the classical treatment is immobilization, analgesic drugs and specific treatment of osteoporosis. However, the immobilization increases demineralization and is potentially harmful, leading to decubitus complications, especially in the elderly.[16]

Vertebral fractures usually cause acute symptoms with pain. Initial symptoms tend to subside after 4 to 6 weeks. However, in some cases the patients have persistent and disabling pain in spite of prolonged medical treatment. In these cases percutaneous vertebroplasty is useful to alleviate symptoms and reduce the duration of immobility.[24,25] Using this method we have treated 40 patients aged

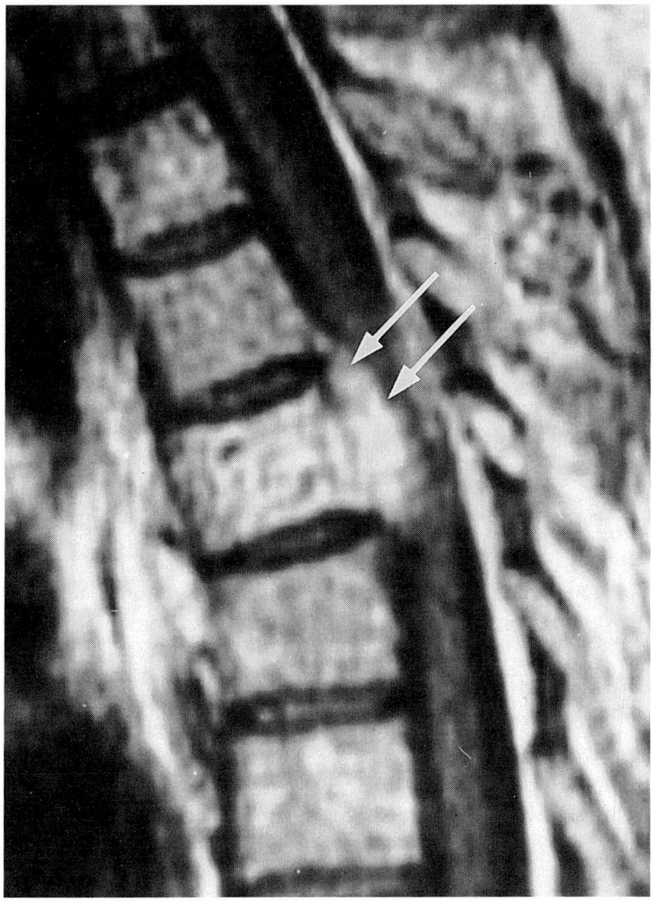

(a)

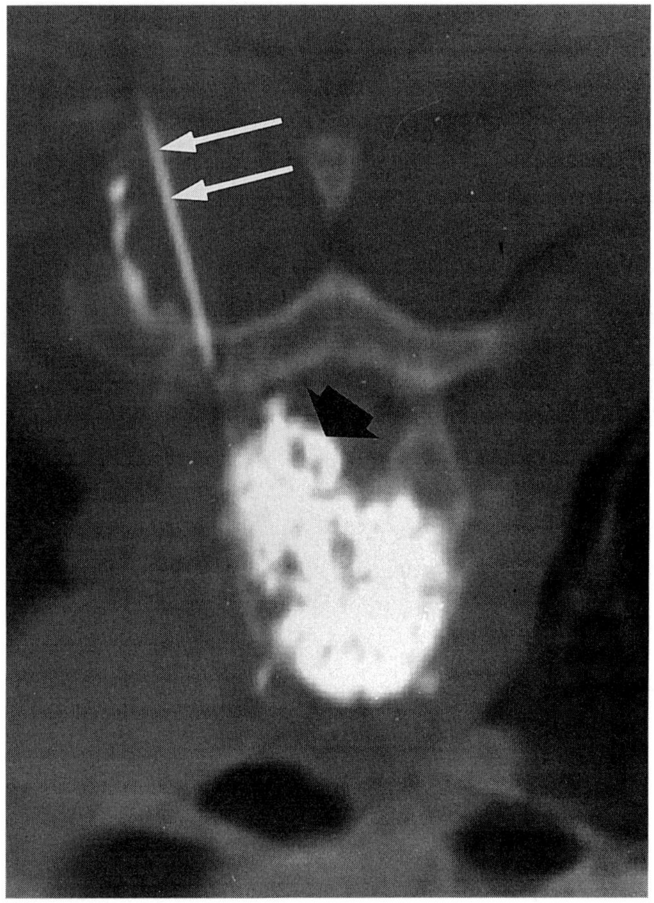

(b)

Fig. 11.9
Thirty-four-year-old man with a T5 aggressive spinal haemangioma with paraparesis. (a) MRI: T1 weighted with gadolinium DTPA shows the epidural component (white arrows). (b) After percutaneous vertebroplasty with acrylic cement, a puncture of the posterior arch with a 20 G needle (white arrows) allows histoacryl embolization of the remaining vertebral haemangioma and of the epidural component (black arrow). The difficulty of surgery was eased by a bloodless operative field.

49–86 years. In 30 cases the treated level was the dorsolumbar junction. Two vertebrae were treated in 13 cases (Fig. 11.6), three vertebrae in 2 cases. All the patients had a vertebral collpase with persistent and disabling pain in spite of immobilization and medical treatment. All patients had previously been treated by analgesic drugs, calcium and immobilization. Because of the persistence of pain, the vertebroplasty was performing between 3 weeks and 5 months after the initial presentation. For all patients, pain was the principle cause of medical consultation and subsequent hospitalization.

A bone biopsy was performed immediately before vertebroplasty in all cases, as part of the same procedure (Figs 11.3 and 11.4).

Results were excellent in all but 1 case, with immediate and complete relief of pain, and the patients were capable of standing up and walking 24–48 hours after the procedure. This effect has proven to be prolonged, and the follow-up (up to 4 years in 10 patients) proves the efficacy of the method (Fig. 11.10). We have seen no serious complications from this technique used in osteoporotic fracture. A potential risk is collapse of the adjacent vertebra in contact with the one injected with cement. We observed this collapse in 1 case and performed a second vertebroplasty at the newly affected level. Vertebroplasty should only be performed when medical treatment has failed. It should be associated with medical treatment of the osteoporosis and sometimes an orthopaedic corset to stabilize the spine.[26]

Spinal tumours

Spinal tumours that may be treated are malignant tumours, metastases, lymphomas or myeloma. (Benign tumours are uncommon indications.) These tumours

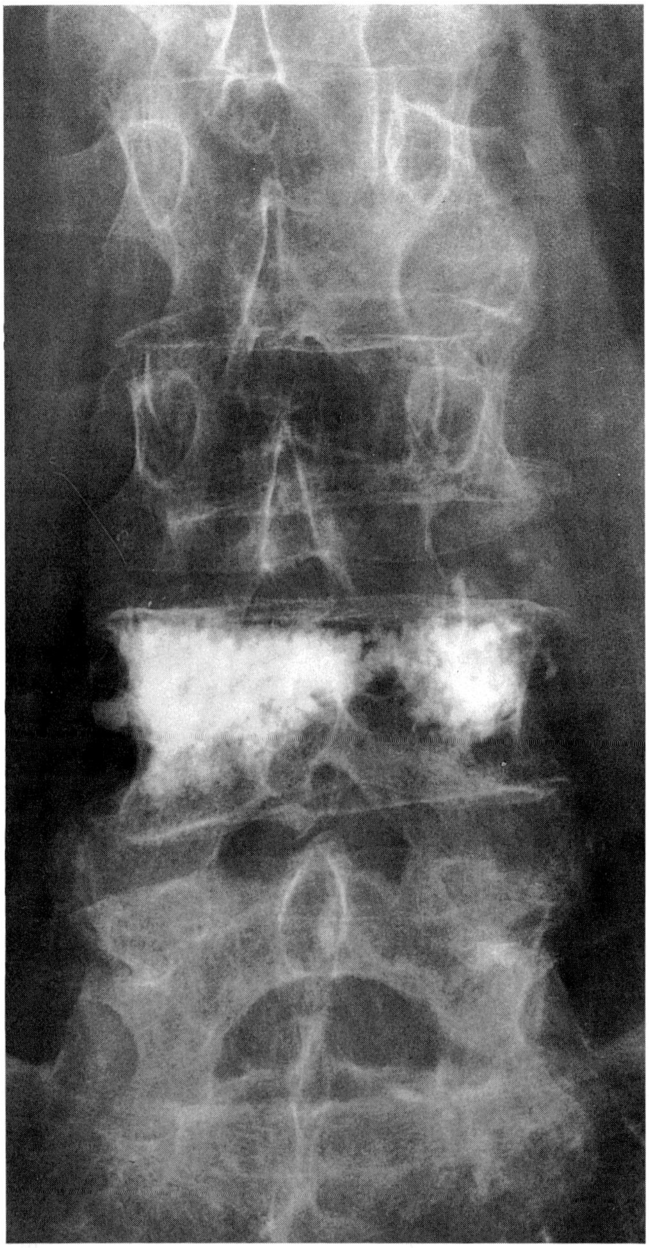

Fig. 11.10
L4 osteoporotic vertebral collapse: 2 years after percutaneous vertebroplasty.

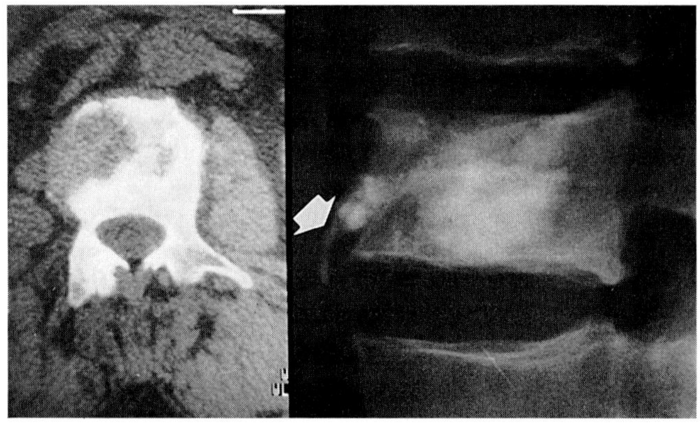

Fig. 11.11 *L3 kidney metastasis*
CT scan before cement injection (on the left). Cementation of a prevertebral vein did not produce symptoms (white arrow).

induce fragmentation of the vertebra and often manifest as spinal pain. Medical pain therapy, surgery, radiotherapy, embolization and chemoembolization are the treatments which could be applied, but each has its failures and complications.[27–33]

Percutaneous intraosseous injection of drugs may induce tumour sclerosis or consolidation.[1,34–36]

We have treated 70 patients (metastases, myelomas, lymphomas) with good results. A good result, which was obtained in about 70 per cent of cases, was judged as significant relief of the pain with return to activities of normal living. Spinal pain was the major symptom in most cases but it was accompanied by radicular pain in a few instances. In 6 cases, spinal cord compression required prior surgery before vertebroplasty was performed.

For the greatest success, the whole vertebral body must be injected even if only a part of it is affected. It is very important to study radiological examinations (plain films, CT and MRI) before injection to detect fracture of the vertebral endplates, intravertebral disc herniations, rupture of the posterior wall and root canal extension of the tumour. This is essential to avoid extravasation of the cement injection. It seems that cement injection reduces tumour development because it acts as an *in situ* embolization. Intra-arterial embolization has the same effect on tumour development. Perhaps ischaemia also reduces pain.[27–29]

What are the main indications in spinal tumours? Theoretically, the best case would be an osteolytic lesion of the vertebral body, with or without vertebral collapse, without rupture of the posterior wall, without radicular or medullary compression, and with intense segmental rachialgia. In practice matters are very different. Each case is unique and must be considered with a clinical and radiological multidisciplinary approach.

In spinal tumours, vertebroplasty is a palliative treatment and not curative. It must be associated with other therapy. For metastases, radiotherapy is very efficient because of hypervascularization and the leucocytic component of the lesion.[37] Hormone therapy and chemotherapy can be useful for primary cancer and in the treatment of metastases.[24,30] Biphosphonate reduces osseous resorption and stops osteoclastic activity. It seems to have an analgesic effect whilst reducing any hypercalcaemia.[38] In myeloma or plasmocytomas, radiotherapy is effective. In malignant lesions, radiotherapy and vertebroplasty should be co-ordinated.

It is logical to perform vertebroplasty before radiotherapy because the analgesic affect occurs immediately after cement injection whilst it is delayed after radiotherapy. There is no modification of the radiotherapy protocols after cement injection and no modification in the cement technique prior to irradiation.[39]

Percutaneous vertebroplasty can be an adjunct to surgery. It can precede or follow surgery in order to bring about a consolidation of the tumoural level, limiting the required strength and extent of orthopaedic devices. It may improve the stability of the surgical fixation.[40,41]

Other substances may be injected into tumoural lesions. Sometimes it is of value to obtain a sclerosis rather than a consolidation or an analgesic effect. In such cases it is possible to inject alcohol or Ethibloc.[35]

The excellent analgesic effect of PVMM has encouraged us to treat other bone lesions outside the spine, metastases and other tumours, especially hip metastases, by a percutaneous approach.[42]

Conclusion

Percutaneous vertebroplasty with acrylic cement is an original and effective therapy. The effects are analgesic and consolidation.

The main indications are the treatment of lesions that fragment the vertebral body: vertebral haemangiomas, osteoporosis and malignant spinal tumours. Other bone lesions, malignant tumours or benign tumours may also be treated by a percutaneous approach. Consolidation and analgesic effects are not always the goal and other substances may be injected to produce sclerosis or consolidation.

The decision to undertake percutaneous vertebroplasty should be the result of a multidisciplinary approach, considering medical treatment in case of osteoporotic collapse, radiotherapy, surgery or chemotherapy in the case of spinal tumours, and surgery in the case of aggressive haemangioma with epidural extension.

References

1 Deramond H, Galibert P, Debussche-Depriester C, Pruvo JP, Heleg A, Hodes JE (1990). Percutaneous vertebroplasty with methyl methacrylate: Techique, method, results. *Radiology* **117**: 32.

2 Deramond H, Debussche-Depriester C, Pruvo JP, Galibert P (1990). La vertébroplastie. *Feuillet de Radiologie* **30**: 262–8.

3 Bonneville JF, Clarisse J, Deramond H (1992). Interventional radiology. In: *Imaging of the Spine and Spinal Cord*, pp. 851–78. Edited by Manelfe C. Raven Press, New York.

4 Gangi A, Kastler BA, Dietman JL (1992). Percutaneous vertebroplasty with CT fluoroscopic guidance. *Radiology* **185**: 194.

5 Bascoulergue Y, Duquesnel J, Leclerq R, Mottolese C, Lapras C (1988). Percutaneous injection of methyl methacrylate in the vertebral body for the treatment of various diseases: Percutaneous vertebroplasty (abstr.), 74 Th RSNA, Chicago.

6 Schlag G, Schliep HJ, Dingledein E, Grieben A, Ringsdore W (1976). Does methyl methacrylate induce cardiovascular complications during alloarthoplasty surgery of the hip joint? *Anaesthesist* **25**: 60–7.

7 Feith R (1975). Side-effects of acrylic cement implanted into bone. *Acta Orthopaedica Scandinavica* **161** (Suppl.): 200–8.

8 Galibert P, Deramond H, Rosat P, Le Gars D (1987). Note préliminaire sur le traitement des angiomes vertébraux par vertébroplastie percutanée. *Neurochirurgie* **33**: 166–8.

9 Darrason R (1988). Place de la vertébroplastie percutanée acrylique dans le traitement des hémangiomes vertébraux aggressifs. MD thesis, Université de Picardie.

10 Deramond H, Darrason R, Galibert P (1989). Percutaneous vertebroplasty with acrylic cement in the treatment of aggressive spinal angiomas. *Rachis* **1**: 143–53.

11 Galibert P, Deramond H (1990). La vertébroplastie acrylique percutanée comme traitement des angiomes vertébraux et des affections dolorigènes et fragilisantes du rachis. *Chirurgie* **116**: 326–35.

12 Fox MW, Onofrio BM (1993). The natural history and management of symptomatic and asymptomatic vertebral hemangiomas. *Journal of Neurosurgery* **78**: 36–45.

13 N'Guyen JP, Djindjian M, Pavlovitch JM, Badian S (1989). Hémangiomes vertébraux avec signes neurologiques: Les resultats thérapeutiques. Enquête de la NF. *Neurochirurgie* **35**: 299–303.

14 Djindjianm, N'Guyen JP, Gaston A, Pavlovitch JM, Poirier J, Awad IA (1992). Multiple vertebral hemangiomas with neurological signs: Case report. *Journal of Neurosurgery* **76**: 1025–8.

15 Heaney RP (1992). The natural history of vertebral osteoporosis: Is low bone mass an epiphenomenon? *Bone* **13** (Suppl. 2): 23–6.

16 Obrant KJ, Bengner U, Johnell O, Nilson BE, Sernbo I (1989). Increasing age-adjusted risk of fragility fractures: A sign of increasing osteoporosis in successive generations? *Calcified Tissue International* **44**: 145–67.

17 Riggs BL, Melton LJ (1988). *Osteoporosis, Etiology, Diagnosis and Management*. Raven Press, New York.

18 Kanis JA, McCloskey EV (1992). Epidemiology of vertebral osteoporosis. *Bone* **13** (Suppl. 2): 1–10.

19 Silverman S (1992). The clinical consequences of vertebral fracture. *Bone* **13** (Suppl. 2): 27–31.

20 Minaire P (1989). Immobilisation osteoporosis: A review. *Clinical Rheumatology* **8** (Suppl. 2): 95–108.

21 Gennari C (1989). Salmon calcitonin (Miacalcic) nasal spray in prevention and treatment of osteoporosis. *Clinical Rheumatology* **8** (Suppl. 2): 61–5.

22 Levernieux J, Julien D, Caulin F (1986). The effect of calcitonin on bone pain and acute resorption related to recent osteoporotic crush fractures: Result of a double

bind and an open study. In: *Calcitropic Hormones and Calcium Metabolism*, pp. 171–8. Edited by Cecchetin M, Segreg G. Elsevier, Amsterdam.

23 Reginster JY (1992). Management of high turnover osteoporosis with calcitonin. *Bone* **13**(Suppl. 2): 37–40.

24 Debussche-Depriester C, Deramond H, Fardellone P, Heleg A, Sebert JL, Cartzl, Galibert P (1991). Percutaneous vertebroplasty with acrylic cement in the treatment of osteoporosis vertebral crush fracture syndrome. *Neuroradiology* **33**(Suppl.): 149–52.

25 Lapras C, Mottolese C, Deruty R, Lapras CH, Remond J, Duquesnel J (1989). Injection percutanée de méthyl méthacrylate dans le traitement de l'ostéoporose et ostéolyse vertébrale grave (technique de P Galibert). *Annals de Chirurgie* **43**(5): 371–6.

26 Parfitt AM (1992). Implications of architecture for the pathogenesis and prevention of vertebral fracture. *Bone* **13**(Suppl. 2): 41–7.

27 Chiras J, Gaston A, Gaveau T, Sellier N, Marsault C, Bories J (1983). Embolisation pré-opératoire en pathologie rachidienne: A propos de 21 cas. *Journal de Radiologie* **64**: 397–402.

28 Meder JF, Ricolfi F, Chiras J, Herbreteau D, Fredy D (1990). Embolisation des tumeurs du rachis. *Feuillets de Radiologie* **30**: 269–73.

29 Courtheoux P, Alachkar F, Casasco A, Adam Y, Derlon JM, Courtheoux F, L'Hirondel JL, Theron J (1985). Chimio-embolisation des métastases du rachis lombaire: Etude préliminaire. *Journal of Neuroradiology* **12**: 151–62.

30 Maxon HR, Schroder LE, Thomas SR, Hertzberg VS, Deutsch EA, Scher HI, Samaratunga RC, Lisbon KF, Williams CC, Moulton JS, Schneider HJ (1990). Re-186 (Sz) HEDP for treatment of painful metastasis: Initial clinical experience in 20 patients with hormone-resistant prostate cancer. *Radiology* **176**: 155–9.

31 Galasko CS (1998). The role of the orthopedic surgeon in the treatment of bone pain. *Cancer Surveys* **7**: 103–25.

32 Nielsen OS, Nunro AJ, Tahnok IF (1991). Bone metastasis: Pathophysiology and management policy. *Journal of Clinical Oncology* **9**: 509–24.

33 Poulsen HS, Nielsen OS, Klee M, Rorth M (1989). Palliative irradiations of bone metastasis. *Cancer Treatment Reviews* **16**: 41–8.

34 Kaermmerlen P, Thiesse P, Bouvar H, Biron P, Monex F, Jonas P (1989). Vertébroplastie percutanée dans le traitement des métastases: Technique et résultats. *Journal of Radiologie* **70**: 557–62.

35 Kaermmerlen P, Thiesse P, Jonas P, Duquesnel J, Bascoulergue Y, Lapras C (1989). Percutaneous injection of orthopedic cement in metastatic vertebral lesion. *New England Journal of Medicine* **321**: 121.

36 Gangi A, Kastler BA, Klinkert A, Mortazani R, Dieteman JL (1992). Percutaneous injection of ethanol: A new method for pain therapy in bone metastasis. *Radiology* **185**: 195.

37 Gilbert HA, Kagam AR, Nussbaum H (1977). Evaluation of radiation therapy for bone metastases: Pain relief and quality of life. *American Journal of Roentgenology* **129**: 1095–6.

38 Powles T (1991). Clodronate: The potential for the future. *Bone* **12**(Suppl. 1): 540–4.

39 Eftekhar MS, Thurston CW (1975). Effect of irradiation on acrylic cement with special reference to fixation of pathological fractures. *Journal of Biomechanics* **8**: 53–6.

40 Harrington KD (1981). The use of methyl methacrylate for vertebral body replacement and anterior stabilisation of the pathological fracture dislocation of the spine due to metastatic malignant disease. *Journal of Bone and Joint Surgery* **63A**: 36–46.

41 Young RF, Post EM, King GA (1980). Treatment of spinal epidural metastasis: Radiological prospective comparison of laminectomy and radiotherapy. *Journal of Neurosurgery* **53**: 741–8.

42 Hamon-Kerautret M, Cotten A, Lejeune JP, Cortet B, Deramond H, Clarisse J (1994). Traitement multidisciplinaire d'un angiome vertébral dorsal agressif. *Revue d'Imagerie Médicale* **5**(12): 769–72.

Index

Note: Page references in *italics* refer to Figures; those in bold refer to Tables.